The Future of Risk Management

Edited by Larisa Ivascu, Marius Pislaru and Lidia Alexa

Published in London, United Kingdom

The Future of Risk Management
http://dx.doi.org/10.5772/intechopen.1001758
Edited by Larisa Ivascu, Marius Pislaru and Lidia Alexa

Contributors
Abdullahi Suleiman Arabi, Adam Suleiman Murtala, Emine Can, Emin Tarakçi, Ewa Kurowska, Hui Zhang, Idris Isa Funtua, Jie Niu, Kiyoshi Nagata, Larisa Ivascu, Lebogang Cleopatra Phama, Lechang Xu, Lidia Alexa, Maasago Mercy Sepadi, Madhurjit Singh Rathore, Marius Pislaru, Monument Thulani Bongani Makhanya, Musa Abdullahi Ali, Musa Suleiman Abdulhamid, Naimeh Borjalilu, Pooran Mal Meena, R.K. Aggarwal, Ramu Meena, Siddhartha Sampath, Valerii Ye Vorotin, Xiangnan Dai, Yalan Wang, Zainab Tukur

First published in London, United Kingdom, 2025 by IntechOpen
IntechOpen is the global imprint of INTECHOPEN LIMITED, registered in England and Wales, registration number: 11086078, 167-169 Great Portland Street, London, W1W 5PF, United Kingdom

For EU product safety concerns: IN TECH d.o.o., Prolaz Marije Krucifikse Kozulić 3, 51000 Rijeka, Croatia, info@intechopen.com or visit our website at intechopen.com.

British Library Cataloguing-in-Publication Data
A catalogue record for this book is available from the British Library

The Future of Risk Management
Edited by Larisa Ivascu, Marius Pislaru and Lidia Alexa
p. cm.

This title is part of the Industrial Engineering and Management Book Series, Volume 9
Topic: Operational Excellence
Series Editor: Fausto Pedro Garcia Marquez
Topic Editor: Stuart So

Print ISBN 978-0-85466-752-9
Online ISBN 978-0-85466-751-2
eBook (PDF) ISBN 978-0-85466-753-6
ISSN 3029-0511

If disposing of this product, please recycle the paper responsibly.

IntechOpen Book Series

Industrial Engineering and Management

Volume 9

Aims and Scope of the Series

Industrial Engineering and Management (IEM) is a discipline that focuses on optimizing complex processes and systems within various industries. It involves the integration of engineering, business, economics, mathematics, and behavioral sciences to improve efficiency, productivity, quality, and overall performance in organizations. Key aspects of Industrial Engineering and Management include: Process Optimization; System Analysis and Design; Quality Control and Management; Supply Chain Management; Operations Management; Human Factors and Ergonomics; Project Management; Cost Analysis and Financial Management; Decision Analysis.

Overall, Industrial Engineering and Management aims to optimize resources, improve processes, enhance productivity, and ensure the effective and efficient utilization of all elements involved in the production or delivery of goods and services. It is crucial in today's competitive business environment for organizations to stay efficient and competitive.

Production Engineering and Operational Excellence are fields of study and practices that focus on optimizing and improving the manufacturing and production processes within an organization. It combines principles from engineering, management, and operational strategies to enhance productivity, efficiency, quality, safety, and sustainability in the production of goods and services.

Here are the key components of Production Engineering and Operational Excellence: Process Optimization; Operational Excellence; Manufacturing Systems Design; Quality Management; Supply Chain Optimization; Production Planning and Scheduling; Automation and Technology Integration; Health, Safety, and Environmental Management; Cost Management; Performance Measurement and Key Performance Indicators (KPIs); Continuous Improvement and Innovation.

Production Engineering and Operational Excellence are crucial for organizations aiming to stay competitive in the global market by achieving high levels of efficiency, quality, and customer satisfaction while optimizing resources and minimizing waste. It is a multidisciplinary approach that encompasses engineering principles, management strategies, and the effective use of technology to drive operational success.

Meet the Series Editor

Fausto Pedro Garcia Marquez is a Full Professor at UCLM, Spain, with accreditation since 2013. He also holds the position of Honorary Senior Research Fellow at Birmingham University, UK, and serves as a Lecturer at the Postgraduate European Institute. In addition to these roles, Fausto has experience as a Senior Manager at Accenture from 2013 to 2014. He earned his European Ph.D. with the highest distinction. Throughout his career, Fausto has received numerous awards and honors. These include the Nominate Prize (2022), Gran Maestre (2022), Grand Prize (2021), Runner Prize (2020), and Advancement Prize (2018), as well as Runner (2015), Advancement (2013), and Silver (2012) by the International Society of Management Science and Engineering Management (ICMSEM). He was also the recipient of the First International Business Ideas Competition 2017 Award. Fausto's contributions extend to academic publishing, with over 242 papers to his name. Notably, his work has been recognized in journals like "Applied Energy" (Q1, IF 9.746, Best Paper 2020) and "Renewable Energy" (Q1, IF 8.001, Best Paper 2014). His affiliations include the editorial and authorship roles in more than 50 books, with publications through respected publishers such as Elsevier, Springer, Pearson, Mc-GrawHill, IntechOpen, IGI, Marcombo, and AlfaOmega. He has authored over 100 international chapters and holds 6 patents. Fausto serves as the Editor of 5 International Journals and is a Committee Member for more than 70 International Conferences. His research portfolio encompasses being the Principal Investigator in 4 European Projects, 8 National Projects, and participating in over 150 projects involving universities and companies. His areas of expertise and research interests span Artificial Intelligence, Maintenance, Management, Renewable Energy, Transport, Advanced Analytics, and Data Science. Fausto is a recognized Expert in the European Union in AI4People (EISMD) and ESF. He also serves as the Director of www.ingeniumgroup.eu, holds the status of Senior Member at IEEE since 2021, and has been honored as an Honorary Member of the Research Council of the Indian Institute of Finance since 2021. Fausto is also the Committee Chair of The International Society for Management Science and Engineering Management (ISMSEM) since 2020.

Meet the Volume Editors

Larisa Ivascu (Ph. D. in Engineering and Management; MBA; B. A. in Software Engineering) is a vice-rector for financial policies and entrepreneurship at Politehnica University of Timisoara in Romania. She is a full-time professor at the Faculty of Management in Production and Transportation, Department of Management, with over 20 years of experience in programming, teaching, and research. Professor Ivascu is the director of the Research Center for Engineering and Management, the president of the Academy of Political Leadership scientific committee, and vice-president of the Society for Ergonomics and Work Environment Management. Professor Ivascu was the head of the Entrepreneurship Office of Politehnica University of Timisoara. She has extensive academic work: published 11 books and 12 chapters in national and international books, coordinated the editing of 7 books at internationally recognized publishing houses, and over 280 academic research studies or articles, and was also a part of international research bodies. Dr. Larisa is a guest editor of important journals, a keynote speaker at various international and national events, part of national and international entrepreneur and academic projects and a visiting professor at prestigious universities.

Marius Pislaru (Ph. D. in Electrical Engineering; M. Sc.; B. S. E.) is a full-time professor at "Gheorghe Asachi" the Technical University of Iasi, Faculty of Industrial Design and Business Management, Department of Engineering and Management, having more than 19 years of experience in business, teaching and research. He is the head of the Engineering and Management Department. Previously, he was a vice-dean responsible for Research, Relations with the Entrepreneurial Environment, and Internationalization at the Faculty of Industrial Design and Business Management. Since 2010, professor Pislaru has been active in the research team of the INTELECTRO company, which was created as an innovative spin-off of the "Gheorghe Asachi" Technical University of Iasi, whose purpose is to carry out the technological transfer of the research results in the industry and to obtain a greater autonomy in the research activity. He has published 7 books, 4 book chapters, and over 150 academic articles. He also has a strong track record of participating in successful national and international projects implemented with European funding on creativity and innovation. He is a member of 7 professional associations and over 20 high-ranking journal reviewers.

Lidia Alexa is currently an Associate Professor in the Engineering and Management Department within the Faculty of Industrial Design and Business Management at "Gheorghe Asachi" Technical University of Iasi and a former Fulbright visiting professor at the AIN Center for Entrepreneurship at Rochester University in New York. She is an experienced professional with over 18 years of experience in entrepreneurship and marketing in Romania's private and public sectors. She has published over 40 scientific papers as an author and co-author in national and international academic journals, focusing on entrepreneurship, entrepreneurship education, open innovation, digital marketing, and social media marketing.

Contents

Preface

This book aims to provide a multidisciplinary approach to risk management. Risk management is essential for competitive organizations and innovation. It aims to develop an effective environment for organizational development by planning, evaluating, analyzing and controlling risks that contribute to achieving organizational objectives. This process is applied in all fields of activity, and the evaluation framework is the same regardless of the field. This volume addresses methods, models, evaluation frameworks, benefits, barriers, and other dimensions of risk management. The targeted risk management directions are digital transformation, marketing risks and strategies and entrepreneurial risks.

Risks are present in all activities and organizational processes. Risk management involves a complex process based on identifying, evaluating, analysing, monitoring and controlling threats to achieve organizational objectives and vision. The complex approach to risks is an important activity and must be carried out in accordance with the interests of the stakeholders and the established objectives. The risk approach must be aligned with organizational policy and development strategies. Treatment techniques are complex and require experience and evaluation of organizational history. The new approaches and the environment dynamics have contributed to the emergence of new risks for which the approach must differ. Each entity must evaluate the cost of the risk, and the measures applied to the consequences must be personalized. Eliminating or minimizing hazards is an important activity. This approach can be found in different areas of activity.

Artificial intelligence (AI) and Machine Learning (ML) are also used in risk assessment. Existing tools can assess risk exposure, risks and their consequences with greater accuracy. There are some sectors of activity where risk assessment and AI implementation are more intense. In the banking sector, chatbots and automatic fraud detection are essential activities. AI and ML tools are used in the front-office and back-office but also to improve the customer experience and satisfaction. In other industries, using these risk management tools has different impact levels. Risk assessment and analysis involve large data sets, as well as the company's history and experience as a risk assessor. In this sense, AI/ML solutions can generate large amounts of accurate and timely data and are thus used to create more effective decisions and strategies. The entire communication process in risk management is improved.

Among the benefits generated by AI/ML in risk management are forecast accuracy, optimization of the variable selection process, identification of hazards, data segmentation and efficient interpretation, ensuring granularity and the development of a coherent basis for the decision-making process, improvement of the decision-making process, efficiency strategies and implementation plans in accordance with the organization's situation. Artificial intelligence is vital in analyzing unstructured data and incomplete data series. Of these benefits, the most important aspect relates to the efficiency and accuracy of risk management strategies. There are also several limitations to using AI/ML in risk assessment. These include high costs, lack of creativity, limited jobs, and reduced communication.

This volume includes a series of topical topics from various fields. Some chapters address innovation, ML models used in risk assessment, occupational risks, or other issues that contribute to the efficiency of organizational processes. As the chapters of the book show, risk management is addressed in ergonomics, soils, business, public management, mines, mechanics, information technology, learning systems, and others. These chapters present different evaluation methods and different approaches. The book is a valuable support for researchers, practitioners, students, professionals, entrepreneurs and employees.

Larisa Ivascu,
Politehnica University of Timisoara,
Timisoara, Romania

Marius Pislaru and Lidia Alexa
"Gheorghe Asachi" Technical University of Iasi,
Iasi, Romania

Section 1

Organizational Innovation

Chapter 1

Managing Digital Transformation Risks in the Context of Organizational Competitiveness

Larisa Ivascu, Marius Pislaru and Lidia Alexa

Abstract

Digitization, digitalization, and digital transformation are increasingly being adopted by organizations across various industries. When talking about digitization, most private companies have passed the initial stages of the process. As these organizations progress through the various stages of digitization, there is a marked shift toward comprehensive digital transformation strategies aimed at maintaining a competitive advantage. The digital transformation of marketing processes is the current approach used by many competitive companies. The marketing activity includes a communication mix that must be innovative and aligned with digitalization. This intense activity of digitalization and digital transformation in marketing has contributed to the emergence of new risks and the adaptation of risk management activities. This chapter aims to identify and understand Romanian entrepreneurs' perceptions of digital transformation of their marketing activities and the associated risks. Using qualitative and quantitative evaluation, the results of the market study are presented. The results show that the main risk is the lack of a clear digitalization strategy, the risk of public defamation (loss of customer trust), and data security concerns. This chapter highlights the impact of digital transformation on organizations, digital marketing, and associated risks. The limitations of the research refer to the fact that the research was carried out in a single country.

Keywords: sustainability, entrepreneurship, marketing, competitive, innovation, risk management

1. Introduction

Marketing is a fundamental component of any organizational strategy and growth as it serves several pivotal functions that collectively enhance the organization's visibility, engagement, and profitability. The main role of marketing is to increase awareness among potential buyers, to involve them in the buying process, and to help them decide to buy. Marketing is a powerful tool that helps entrepreneurs develop a marketing mix adapted to the requirement, create brand awareness, obtain the expected financial results, attract new customers, retain current customers, capture attention, and achieve success. All marketing activities involve a mix between art and science [1–6].

The development of a marketing mix adapted to market requirements involves extensive market research and the identification of the particularities that differentiate the marketing mix from that of the competition. This entails a series of complex activities and risks. Marketing people get involved in using new tools to identify the trends and the needs and desires of their customers. Creating brand awareness is the main stage in marketing activities. Achieving success means developing a strong brand. The main question that must be addressed is "why would customers choose the brand?". If the answer is clear and well-defined, then the choice of the clients is an adequate one. Financial results are important for every entrepreneur. Profit is one of the indicators controlled by shareholders. If the marketing mix can generate a profit, then the marketing activity has been carried out according to the plan. If the results are not as expected, then the product design and the development of the marketing mix must be resumed. Retention is another important dimension. When the entrepreneur manages to segment the market well and cater to the needs and expectations of the target market, then the possibility of customers leaving for the competition is small. Attracting new customers is done if the marketing mix is more attractive than the competition to capture attention. Capturing the attention of customers is done through a good targeting of the market and the involvement of an innovative communication mix. New innovative technologies contribute to the development of innovative marketing. The use of appropriate knowledge is important and complementary to new technologies [1, 6–11].

Digital transformation has led to the emergence of new marketing activities, new activities in the environment and business, and implicitly new risks. Many studies evaluate the impact of digitization on organizations but do not consider a comprehensive approach to market impact in conjunction with marketing and associated risks. This research fills this gap and provides a complete picture. Innovative marketing involves new marketing techniques, new technologies and new impact control and monitoring tools. All these aspects are strategic directions for developing the innovative marketing customers expect. This approach, in which technology is an integral part, involves a series of new risks that must be evaluated and addressed at the organizational level [7, 8].

This research answers the following questions:

1. What are the effects of digital transformation in organizational marketing?

2. What is the perception of entrepreneurs on DT and the organizational risks generated by DT?

This chapter aims to identify and understand Romanian entrepreneurs' perceptions of the digital transformation of their marketing activities and the associated risks. This chapter advances the specialty literature with comprehensive research conducted on new technologies used in marketing activities, sustainability, and technological risks. It aims to identify innovative directions of organizations and technologies that contribute to digital transformation (DT).

2. The evaluation of digital transformation in marketing

2.1 The digital marketing tools

The tools used in digital marketing activity are multiple. They are used by companies depending on their needs, knowledge, market segments, and marketing mix.

The technological evolution led to the rapid evolution of marketing activity as well. The following table shows some tools used in digital marketing (**Table 1**) [11, 12].

Marketing innovation is important for the sustainable development of the organization [12–14]. There are four major categories of innovation depending on technology and market (including customers):

1. Architectural: Existing technologies are used to attract new customers (existing technologies, new customers)

2. Disruptive: Create technologies that are adapted to customer needs (new technologies, new customers)

3. Incremental: The activity through which new products are created or updated using existing technologies (existing technologies, current customers)

4. Radical: Uses new technologies to create new experiences for existing customers. Processes are improved, and products are updated (new technologies, existing customers).

Technology	Description	Marketing implication
Social media marketing	Include influencer marketing. Through this activity, video content, blogs, and posts are developed to create value and develop new opportunities to increase market share.	It creates value for the communication mix and can develop a series of risks associated with the organization's image. The importance of affiliate marketing is constantly growing.
Advertising	It puts the product in a favorable light and increases the interest of customers.	It is the activity that creates the connection between the product and the customer, being a very important one.
Content marketing	Develop content that can be attractive to individuals.	Outlines the importance of products in certain segments.
Email marketing	It informs the population about an existing product.	It creates a connection between the product and the customer.
Search Engine Optimization (SEO)	Optimize web content	It optimizes web content and contributes to improving the visibility of a product in the search process.
Pay-per-click (PPC) advertising	An indicator for accessing a website.	It monitors the number of possible buyers of a website.
Virtual reality	The ability to create unique moments, the possibility to enrich the commitment of customers without products, and to develop a connection between them.	The possibility of developing a customized product that offers customers the opportunity to test the desired product.
Marketing Analytics	It offers the opportunity to analyze campaigns, actual activities and evaluate return on investment (ROI).	It contributes to the control and monitoring of the achievement of the profit level planned by the shareholders.
Mobile marketing	Advertising activity using the mobile phone.	It is a subset of mobile advertising and is appreciated in the new technological dynamics.

Table 1.
Digital marketing tools and their implications.

2.2 The innovative marketing methods

Innovation in marketing is important because it considers the entire business and develops the framework for improving competitiveness. It considers current and future target markets and contributes to the design of unique products, improving customer satisfaction and positioning in the market. Among the innovative marketing methods are the ones presented in **Table 2** [14–16].

All these innovative aspects must be included in the company's strategic plan [17–19]. Aligning the vision, mission, and strategic objectives with the innovative character contributes to the company's sustainable development. The strategy plan for digital marketing includes a series of stages:

1. Establishing the organization's goals to achieve the organization's vision and mission.

2. Prioritization of activities for the interface with clients. Establishing the budget.

3. Identifying innovative tools that match the natural points.

4. Selecting the appropriate tools for the established budget.

5. Evaluation of customer reviews and their wishes.

6. Ensuring scalability for the future.

7. Ensuring the integration of the selected instruments with the existing ones.

8. Implementing the tools in the current process.

Method	Description
Retain	Innovative marketing involves activities to retain customers. The use of different promotion strategies will contribute to customer retention.
Rebrand	It involves activities to modify the brand's logo, packaging, communication, target market, or mission based on changes in the market.
Collaborate	Partnerships and collaboration between organizations are among the important methods of innovative marketing.
Expand	Launching new markets and accessing new opportunities are important methods.
Interact	Interaction with customers is important for developing a collaborative relationship. The realization of events, and the host role, are among the important methods of the current dynamics.
Educate	Developing an appropriate market-to-marketing mix is a very important step. Educating consumers and teaching them about the product's functionalities are essential elements.
Personalize	Developing an enjoyable experience is important for customers. Creating memorable moments for customers is essential.
Partnership	Improving the level of transparency is essential in business activity.

Table 2.
The innovative marketing methods.

9. Compatibility testing and customer response from a quality perspective.

10. Permanent updating according to industry trends.

The advantages of digital marketing tools are important for organizations. Identifying the advantages at the organizational level is important for the interested parties [14, 15]. Among them are the following:

- Cost efficiency: Many of the activities will no longer consume many resources.

- Precise targeting: Digital tools offer the possibility of more accurate targeting, which will contribute in a real way to the level of profit.

- Quick analytical situations: These situations can be obtained in real time and in this way, decisions can be made more easily.

- Accessibility: There is a permanent possibility of obtaining necessary information (24/7).

- Flexible scaling: Any developed concept can be scaled more simply.

- Reduction of losses: Through an improvement of knowledge, reductions of losses in the flow of the process can be obtained.

- Automation: Many of the activities carried out can be automated.

- Personalization: Improving the degree of personalization by using appropriate tools.

Among the disadvantages of the digital and involved transformation of digital marketing are [12, 20]:

- Process complexity: The increased complexity may require new skills and knowledge.

- Lack of standardization: Standardized activities cannot be used in this marketing activity.

- High costs: New technologies cost more and may involve a greater effort on the part of the organization.

- Risk: Different risks that can be developed by technology or by the lack of attractiveness of digital marketing among clients.

- Staff barriers: Often the staff is not willing to learn causing interruptions in the organizational flow or delays.

- Decrease in customer confidence: If the products require knowledge from the customers, they can refuse the products.

- Data security: New technologies can develop certain security risks.

3. The principles of risk management

Digital transformation contributes to the achievement of organizational objectives. At the organization's marketing function level, a series of risks develop that must be planned, identified, evaluated, analyzed, monitored, and controlled. The identification, evaluation, and analysis methods are applied according to the organization's needs. Risk management is an important process in the current conditions of digital transformation. This process involves a series of stages from the identification of risks to their treatment and control. The treatment step of digital marketing risks can include transferring, reducing, avoiding, or retaining them (**Figure 1**) [1, 4, 5, 10, 21].

4. Methodology

The research methodology involves qualitative and quantitative methods for highlighting the results obtained. That is why they must be identified and evaluated. The experience of risk assessors can influence the risk plan. This research used qualitative and quantitative evaluation. The primary data obtained were quantitatively modeled, and the specialty literature was evaluated using qualitative evaluation. To identify and understand the entrepreneurs' perception of digital transformation and the associated risks, market research was carried out. The research was carried out using an online questionnaire on a database of 150 companies from Romania. The response rate was 81%. The valid answers are presented below. Statistical processing and multiple regression were used to highlight the entrepreneurs' perception of the research subject. The results of the research are presented below.

5. Results and discussion

From the perspective of innovation and digital transformation, many entrepreneurs know and are familiar with the principles (**Figure 2**). Among the respondents, more than 100 individuals are above the average limit of the appreciation scale.

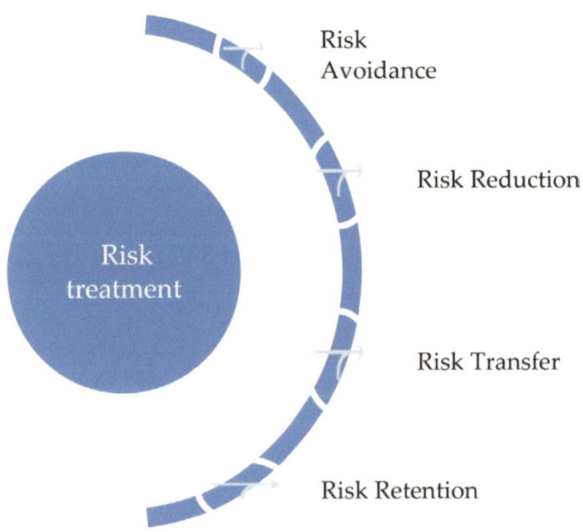

Figure 1.
The risk treatment options for digital marketing.

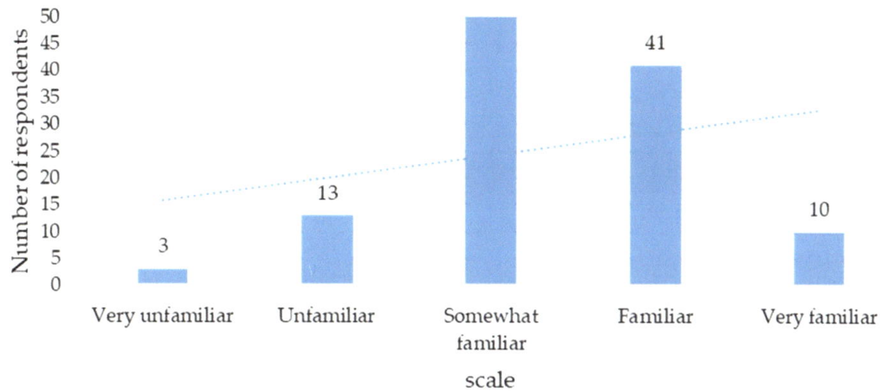

Figure 2.
What is the awareness level of entrepreneurs about innovation and digital transformation?

Information about digital transformation (**Figure 3**), the possibility of financing, and other aspects sometimes reach entrepreneurs later. From this perspective, they appreciated this level of information. The majority show that they are informed and are familiar with some aspects.

Some sectors of activity are more digitized than others. That's why much of the information that shows the answers received comes from the industry and not necessarily from the activity sector (**Figure 4**).

Digital marketing tools are numerous and vary depending on the field of activity. **Figure 5** shows that advertising and marketing analytics and social media marketing are most used in marketing activities.

The evaluation of the answers received reveals that the impact of digital transformation is launched, especially in the emergence of new risks (**Figure 6**). The organization's automation is also evident following the application of digitalization tools and methods.

Data regarding the order of priority of the four categories from the entrepreneurs' perspective is presented in **Figure 7**. Entrepreneurs consider all categories as important. If we examine the average values, we can see that they are equally distributed and underline the importance of all categories.

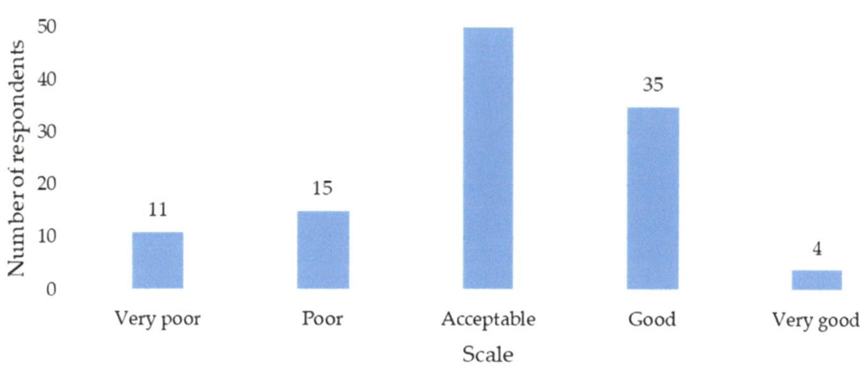

Figure 3.
Information about digital transformation in the entrepreneurial environment.

	Have you come across the term innovation and digital transformation in your business sector?		Have you come across the term innovation and digital transformation in you're the industry?	
	Frequency	Percent	Frequency	Percent
No	41	33.9	65	53.7
Yes	80	66.1	56	46.3
Total	121	100	121	100

Figure 4.
How often do you hear innovation and digital transformation terms in your business sector or industry?

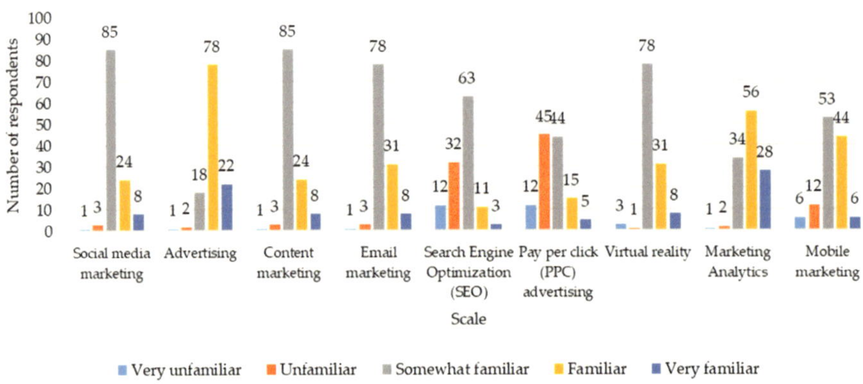

Figure 5.
The use of digital marketing tools by entrepreneurs in organizational activity.

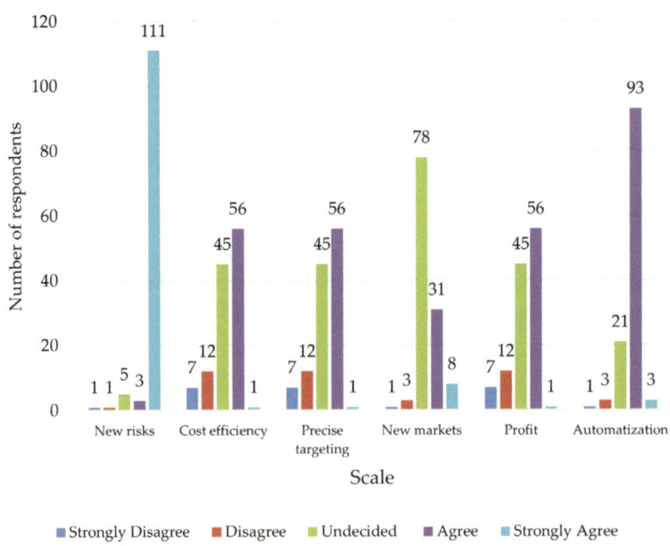

Figure 6.
The impact of digital transformation.

		Green marketing importance	Branding importance	Evolution importance	Education importance
N	Valid	121	121	121	121
Mean		4.52	4.16	4.26	4.58
Median		5.00	5.00	5.00	5.00

Figure 7.
The digital transformation categories.

The entrepreneurs were presented with digital marketing tools to express their opinions and whether or not they used them. Most prefer "social media marketing", "content marketing" and "email marketing" (**Figure 8**). On the opposite end in terms of usage in marketing activities is "mobile marketing".

Using Pearson correlation, we investigated if there is any correlation between gender, age, and the proposed digital transformation tools (**Figure 9**). We found a significant positive relationship between gender and the following tools: advertising, SEO, and mobile marketing. This finding suggests that female entrepreneurs demonstrate a higher propensity to adopt these digital marketing techniques compared to their male counterparts. The results also underscore the importance of considering gender differences when implementing digital transformation strategies, as female entrepreneurs may exhibit distinct preferences and behaviors in their digital marketing approaches.

From the perspective of regression analyses, it was desired to identify potential relationships between variables. The established independent variables were age and gender (**Figure 10**). We wanted to identify the degree of influence of the independent variables on digital transformation. The regression analysis shows that there is no statistically significant relationship between gender and the extent to which digital transformation tools are utilized within organizations. However, the analysis revealed that age significantly influences the use of digital transformation tools. This relationship indicates that as individuals' age varies, so does their propensity to adopt and

	"I do not use it"	"I am indifferent"	"I use it"	Mean
Social media marketing (1)	4	10	107	1.85
Advertising (2)	2	16	103	1.83
Content marketing (3)	3	16	102	1.82
Email marketing (4)	2	24	95	1.77
SEO (5)	4	23	94	1.74
PPC advertising (6)	4	24	93	1.74
Virtual reality (7)	4	24	93	1.74
Marketing Analytics (8)	2	29	90	1.73
Mobile marketing (9)	1	34	86	1.7

Figure 8.
The importance of digital transformation.

		-1	-2	-3	-4	-5	-6	-7	-8	-9
Gender	PC*	0.126	0.179	0.136	0.113	0.19	0.238	0.207	0.205	0.233
	Sig.	0.169	0.049	0.138	0.218	0.037	0.008	0.023	0.024	0.01
Age	PC*	0.074	-0.014	0.097	0.202	0.06	0.107	0.107	-0.097	-0.046
	Sig.	0.418	0.881	0.291	0.026	0.513	0.241	0.241	0.29	0.616

Figure 9.
The correlation between digital transformation tools and demographic data.

Model	Unstandardized Coefficients		Standardized Coefficients	t	Sig.
	B	Std. Error	Beta		
(Constant)	2.019	0.243		7.791	0
Age	.599	.186	.383	3.223	.002
Gender	-0.246	0.191	-0.106	-1.286	0.201

Figure 10.
Multiple regression analysis of digital transformation tools and demographics.

	Severity of Consequence				
	minor	significant	serious	major	fatality
Loss of customer trust	0	1	34	78	8
Data security concerns	0	1	34	78	8
Disruption to employees	4	10	45	54	8
Risk of failure	7	13	41	54	6
Leadership risk	0	1	34	78	8
Clear digitalization strategy	4	10	45	54	8
Financial risks	0	1	33	33	54

	Likelihood of occurrence				
	very unlikely	slight	feasible	likely	very likely
Loss of customer trust	0	1	34	78	8
Data security concerns	0	1	34	78	8
Disruption to employees	1	4	6	43	67
Risk of failure	1	43	56	12	9
Leadership risk	0	1	34	78	8
Clear digitalization strategy	0	4	38	56	23
Financial risks	0	1	33	33	54

Figure 11.
The risk management of innovation marketing.

integrate digital transformation strategies. The significance of this finding can be attributed to the broader context of technological evolution, which greatly affects the adoption rates of new technologies. Younger individuals, often more technologically adept, may be more inclined to leverage digital transformation tools than older individuals.

Figure 11 has two parts in which the severity of the consequences and the likelihood of risk occurrence are evaluated. Seven risks are evaluated: loss of customer trust, data security concerns, disruption to employees, risk of failure, leadership risk, clear digitalization strategy, and financial risks. The results show that the main risk is the lack of a clear digitalization strategy, the risk of public defamation (loss of customer trust), and data security concerns.

6. Conclusions and future direction

The massive digitization of organizations has led to the emergence of new and complex risks. Incorporating technologies into organizational processes has become a priority for all industries and has led to important changes in terms of efficiency, agility, innovation, and the unlocking of organizational values. This approach is present in all fields, being intensively addressed and debated by researchers, teachers, and practitioners. Furthermore, the rapid pace of technological advancements demands continuous adaptation and proactive strategies to mitigate potential threats and capitalize on new opportunities. In this context, risk management is a mandatory and important process for technological development.

The evaluation of new technologies for innovative marketing is very important. Technological development means that this activity is also permanently updated. New technologies also bring with them a series of risks. The risk assessment from this research shows that the main risks perceived by entrepreneurs are the lack of a clear digitalization strategy, the risk of public defamation (loss of customer trust), and data security concerns.

The answers to the research questions highlight an alignment of the organization's needs with the evolution of digitalization.

From the perspective of the effects of digital transformation on organizational marketing, marketing takes on new, more dynamic forms, which have a direct impact on the client or user. Many of the marketing actions are automated, and the input of the marketing man takes new forms.

Entrepreneurs' perception of DT and the organizational risks generated by DT highlight an acceptance of technological progress and the use of new tools for the promotion of organizations. Risk consequences require a risk-hazard approach to minimize the severity of consequences. Correctly approached risk management contributes to improving the innovation of the marketing activity.

As markets and industries evolve, clients and consumers are changing their behaviors and decision-making processes, and this shift introduces new market risks and opportunities for innovative marketing strategies. At the same time, the massive development of organizations in the online environment has contributed to a series of cyber risks that must be evaluated at the organizational level. All these aspects imply the development of organizational strategies in which risk management and sustainability should be a priority. Any attempt to access a new market involves assuming a certain risk. Therefore, organizations must adapt by implementing comprehensive risk management practices and ensuring sustainable operations to thrive in this dynamic landscape.

Conflict of interest

All authors declare that they have no conflicts of interest.

Author details

Larisa Ivascu[1,2]*, Marius Pislaru[3] and Lidia Alexa[3]

1 Politehnica University of Timisoara, Timisoara, Romania

2 The Academy of Romanian Scientists, Bucharest, Romania

3 "Gheorghe Asachi" Technical University of Iasi, Romania

*Address all correspondence to: larisa.ivascu@upt.ro

References

[1] Syrová L, Špička J. Exploring the indirect links between enterprise risk management and the financial performance of SMEs. Risk Management. 2023;**25**(1):1-27. DOI: 10.1057/s41283-022-00107-9

[2] Borysenko O, Vasylyeva O, Katerna O, Masiuk I, Panakhi O. Modeling the risks of the global customs space. Journal of Risk and Financial Management. 2022;**15**(12):598. DOI: 10.3390/jrfm15120598

[3] Schuett J. Risk management in the artificial intelligence act. European Journal of Risk Regulation. 2023;**15**(2):367-385. DOI: 10.1017/err.2023.1

[4] Tziakou E, Fragkaki AG, Platis A. Identifying risk management challenges in laboratories. Accreditation and Quality Assurance. 2023;**28**(4):167-179. DOI: 10.1007/s00769-023-01540-3

[5] Buganová K, Hudáková M, Šimíčková J, Mošková E. Disparities in the implementation of risk management in the SMEs. Systems. 2023;**11**(2):71. DOI: 10.3390/systems11020071

[6] Shubailat OM, Al-Zaqeba MAA, Madi A, Ababneh AM. Customs intelligence and risk management in sustainable supply chain for general customs department logistics. Uncertain Supply Chain Management. 2024;**12**(1):387-398. DOI: 10.5267/j.uscm.2023.9.013

[7] Ivascu L. Green marketing strategies and market barriers to sustainability products design. In: Proceedings of 2018 the 8th International Workshop on Computer Science and Engineering, WCSE 2018. 2018. DOI: 10.18178/wcse.2018.06.138

[8] Ivascu L. Education for Sustainability: Current Status, Prospects, and Directions. 2017. DOI: 10.15405/epsbs.2017.01.02.27

[9] Lup PS, Ivascu L, Prostean G. Manipulation and risk in the TV commercial at sweets for childrens. In: Proceedings of the 31st International Business Information Management Association Conference, IBIMA 2018: Innovation Management and Education Excellence through Vision 2020. 2018

[10] Ivascu L, Cioca LI, Tamasila M, Taucean I, Bigan C. Risk management and an approach for introducing functional safety in the development of automotive E/E systems. In: Proceedings of the 29th International Business Information Management Association Conference - Education Excellence and Innovation Management through Vision 2020: From Regional Development Sustainability to Global Economic Growth. USA: International Business Information Management Association-IBIMA; 2017

[11] Ivascu L, Cioca L-I, Izvercian M. Investigating the relationship between risk management and sustainable development management within the organizations. In: Proceedings of the 11th International Conference on Innovation and Management. Vol. I and II. Vaasa, Finland: Wuhan University Technology Press; 2014

[12] Munir AR, Kadir N, Umar F, Lyas GB. The impact of digital marketing and brand articulating capability for enhancing marketing capability. International Journal of Data and Network Science. 2023;**7**(1):65-72. DOI: 10.5267/j.ijdns.2022.12.005

[13] Apasrawirote D, Yawised K, Muneesawang P. Digital Marketing

Capability: The Mystery of Business Capabilities. 2022. DOI: 10.1108/MIP-11-2021-0399

[14] Taherdoost H. Digital marketing. In: EAI/Springer Innovations in Communication and Computing. Vol. Part F1354. 2023. DOI: 10.1007/978-3-031-39626-7_9

[15] Masrianto A, Hartoyo H, Hubeis AVS, Hasanah N. Digital marketing utilization index for evaluating and improving company digital marketing capability. Journal of Open Innovation: Technology, Market, and Complexity. 2022;**8**(3):153. DOI: 10.3390/joitmc8030153

[16] Rizvanović B, Zutshi A, Grilo A, Nodehi T. Linking the potentials of extended digital marketing impact and start-up growth: Developing a macro-dynamic framework of start-up growth drivers supported by digital marketing. Technological Forecasting and Social Change. 2023;**186**:122128. DOI: 10.1016/j.techfore.2022.122128

[17] Turi A, Mocan M, Ivascu L, Goncalves G. Managing success: The benefits of marketing effectiveness and strategic plan adaptation in automotive industry. In: Proceedings of the 29th International Business Information Management Association Conference - Education Excellence and Innovation Management through Vision 2020: From Regional Development Sustainability to Global Economic Growth. USA: International Business Information Management Association-IBIMA; 2017

[18] Dragan F, Ivascu L, Ardelean B-O. Trends for Manufacturing Industry: A Strategic Roadmap toward Industry 5.0. 2022. DOI: 10.1007/978-981-16-7365-8_11

[19] Bertea OO, Coroian A-M, Lăcraru AA, Popescu I, Ivaşcu L. Strategic management in small businesses during Covid-19. Review of Management and Economic Engineering. 2021;**20**(3):2360-2155

[20] Alexa L, Apetrei A, Pîslaru M. Fast fashion—An industry at the intersection of green marketing with greenwashing. In: International Symposium "Technical Textiles - Present and Future". 2022. DOI: 10.2478/9788366675735-042

[21] Fedotov O, Zotenko O. Risk management in customs inspection of goods. Lex Portus. 2020;**3**(23):79-94. DOI: 10.26886/2524-101X.3.2020.5

Chapter 2

Establishing Information Security Policy as an Organizational Risk Management

Kiyoshi Nagata

Abstract

In the advanced information and communication network society, every organization faces information-related risks such as information leaks, system, and service malfunctions, unauthorized intrusions, business email compromise, ransom attacks, etc. In order to deal with these various types of risks, it is necessary to take measures that emphasize the balance of the entire organization rather than individual technical measures. In this chapter, we will provide an overview of various risks related to information and consider the establishment of information security policies as a means of overall risk management. Especially keeping in mind SMEs with limited financial and human resources, we will discuss the information security policy automatic generation system by utilizing ontology.

Keywords: information related risks, information security policy, SMEs, automatic generation system, ontology

1. Introduction

World Economic Forum analyzes the risks to the global economy and the world as a whole and publishes the results every year in "The Global Risks Report" underpinned by the Global Risks Perception Survey (GRPS). Responses for the GRPS 2022–2023 were collected from September 7 to October 5, 2022, and have brought together leading insights on the evolving global risks landscape from over 1200 experts across academia, business, government, international communities, and civil society.

Global risk is defined as the possibility of the occurrence of an event or condition which, if it occurs, would negatively impact a significant proportion of global GDP, population, or natural resources. The risk of "Widespread cyber-crime and cyber insecurity" which is described as increasingly sophisticated cyberespionage or cybercrimes including loss of privacy, data fraud or theft, and cyber espionage is ranked 8th risk in the 18th Edition of the report ([1], p. 8).

It also reported that there were some notable differences between the responses of government and business respondents, with "Debt crises", "Failure to stabilize price trajectories", "Failure to mitigate climate change", and "Failure of climate change adaptation" featuring more prominently for governments, and "Widespread cybercrime and cyber insecurity", and "Large-scale environmental damage incidents" featuring higher for business.

From the response by asking, "Please rank the top 5 currently manifesting risks in order of how severe you believe their impact will be on a global level in 2023", the "Cyberattacks on critical infrastructure" is the 5th risk as the currently manifesting risks, following the "Energy supply crisis (1st)", "Cost-of-living crisis(2nd)", "Risking inflation(3rd)", and "Food supply crisis(4th)". As highlighted in the chapter 'Digital Dependencies and Cyber Vulnerabilities' in Global Risks Report 2022, malicious activity in cyberspace is growing, with more aggressive and sophisticated attacks taking advantage of more widespread exposure ([2], pp. 45–56). It was seen as a persistent threat by GRPS respondents as well as a strong driver of other risks. Therefore, "Widespread cybercrime and cyber insecurity" is considered a new entrant into the top 10 rankings of the most severe risks over the next decade.

In the reports, examples of three concerned risks are cited with descriptions as follows ([1], p. 76):

- Breakdown of critical information infrastructure: Deterioration, overload, or shutdown of critical physical and digital infrastructure or services leading to the breakdown of internet, cellular devices, public utilities, or satellites. Those are stemming from, but not limited to, cyberattacks, intentional or unintentional physical damage, or solar storms.

- Digital inequality and lack of access to digital services: Fractured or unequal access to digital networks and technologies stemming from underinvestment, low digital skills, insufficient purchasing power, or government restrictions on technologies.

- Digital power concentration: Concentration of critical digital assets, capabilities, or knowledge among a small number of individuals, businesses, or states that can control access to digital technologies and demand discretionary pricing. Those are stemming from, but not limited to, the failure of anti-trust regulation, inadequate investment in the innovation ecosystem, or state control over key technologies.

These digital risks are considered critical risks, especially in developing countries and oil-producing countries. Digital inequality is the first-ranked risk in the executive opinion survey in India and the fourth-ranked risk in Indonesia. The digital power concentration is the fifth among those in China. Executives in European countries such as Austria, Denmark, Hungary, Poland, and Switzerland recognize the breakdown of critical infrastructure through cyber attacks as one of the five critical risks.

Statista, a global data and business intelligence platform, takes the following as examples of cybercrime: Identity fraud, data theft, ransomware attacks, copyright infringement, and phishing campaigns. From their research, the most common type of cyber attack that organizations worldwide experienced in 2022 was bulk phishing, and SMS phishing, or smishing, and ransomware were quite common too. Statista also estimates the cost of cybercrime worldwide at $8.44 trillion in 2022 and $11.5 trillion in 2023. The predicted amount of cybercrime will rise to $23.82 trillion in 2027 [3]. eSentire, globally recognized as the Authority in Managed Detection and Response (MDR), also predicted that the global annual cost of cybercrime will reach $8 trillion in 2023, according to Cybersecurity Ventures, and is expected to reach $10.5 trillion by 2025 [4].

A Statista report states that organizations worldwide don't only pay to get the lost data back but also suffer downtime and disruption in operations caused by cybercrime. The average cost of a data breach worldwide is around $4.35 million, but financial repercussions differ greatly depending on the region, organization size, and industry. The average cost of a data breach in the healthcare sector is $10.1 million.

FBI (Federal Bureau of Investigation)'s Internet Crime Complaint Center (IC3) collects reports on cyberattacks and incidents from U.S. residents, and publishes the result of analyzed data, identifying trends, and pursuing the threat at hand in the "FBI Internet Crime Report". In 2022, the IC3 received 800,944 complaints with a potential total loss of more than $10.3 billion in 2022, which has grown from $6.9 billion in 2021 ([5], p. 7). Here let us list some characteristic risks from the report in 2022 ([5], pp. 11–16).

- BUSINESS EMAIL COMPROMISE (BEC): 21,832 BEC complaints received by IC3 were adjusted losses of over $2.7 billion. BEC is a sophisticated scam that is frequently carried out when a subject compromises legitimate business email accounts through social engineering or computer intrusion techniques to conduct unauthorized transfers of funds. BEC targets not only businesses but also individuals performing transfers of funds.

- INVESTMENT (Fraud Losses): Investment scams reported to the IC3 were the costliest scheme. The fraud losses increased from $1.45 billion in 2021 to $3.31 billion in 2022 within which cryptocurrency investment fraud rose from $907 million to $2.57 billion. The covering losses from these fraudulent investments are assumed massive debt, and the most targeted age are from 30 to 49.

- RANSOMWARE: 2385 complaints identified as ransomware received by IC3 have adjusted losses of more than $34.3 million. Ransomware is a type of malicious software that encrypts data on a computer, making it unusable. Until the ransom is paid, the cyber-criminal will often steal data off the system and hold that data hostage. If the ransom is not paid, the victim's data remains unavailable.

- CALL CENTER FRAUD: TECH AND CUSTOMER SUPPORT/GOVERNMENT IMPERSONATION: Illegal call center defraud was classified into two categories Tech/Customer Support and Government Impersonation. They are responsible for over $1 billion in losses to victims.

As we see above, in an advanced information and communication network society, ensuring the security of information is the basis of organizational activities, and any act that damages them poses a great risk to the sustainable operation of the organization. Prior to the beginning of commercial use of the Internet in the early 1990s, information security in organizations was largely handled by core information processing departments and was a technical and tactical approach by experts. Currently, it is an organizational and strategic approach involving all stakeholders, including employees, and must take into account the three elements of information security: Confidentiality, Integrity, and Availability (CIA) in a well-balanced manner.

The International Institute for Management Development (IMD), one of the world's leading business schools based in Switzerland, publishes the IMD Word Digital Competitiveness Ranking every year [6]. In a recent report, they insist that cybersecurity capabilities, both at the company and governmental level, have become a very important factor, and the result reflects those factors facilitating the strengthening of capabilities to protect digital infrastructure from cyber-attacks [7]. The ranking evaluation is based on the three main digital competitiveness factors, such as "Knowledge", "Technology", and "Future Readiness", and each of them has three subfactor categories to which all the evaluation factor items belong.

Among 40 countries with GDP per capita, greater than $20,000, countries or regions ranked in the top 20 are listed in **Table 1** according to the evaluation item "Cyber security" along with some other evaluation item rankings.

From the Appendices and Sources in ([6], pp. 182–184), short descriptions of each item are as follows;

- Cyber security: Cyber security is being adequately addressed by corporations.
- Government cyber security capacity: The government's capability to mitigate harm from cyber security threats.
- Digital/technology skills: Digital/technological skills are readily available.
- Opportunities and threats: Companies are very good at responding quickly to opportunities and threats.

Country or region	Cyber security	Government cyber security capability	Digital/ Technology skills	Opportunity and threats	Agility of companies
Qatar	1	13	11	11	23
Saudi Arabia	2	21	7	20	20
Finland	3	34	3	15	15
Austria	4	26	40	22	14
UAE	5	7	16	23	8
Singapore	6	10	9	15	10
Hong Kong SAR	7	48	15	2	4
Israel	8	1	19	24	24
Taiwan	9	9	33	5	3
China	10	3	12	13	22
Canada	12	4	14	18	18
Sweden	13	17	4	7	7
Denmark	14	8	5	1	1
Switzerland	15	27	18	8	9
Estonia	16	2	44	28	11
Iceland	17	52	1	3	2
Norway	18	44	8	10	13
Lithuania	19	32	2	4	5
Netherlands	20	40	6	9	12
USA	27	15	10	19	21

The ranking value is among 64 countries or regions.
Source: IMD Word Digital Competitiveness ranking 2023.

Table 1.
Top 20 Countries or Regions of GDP per capita greater than $20,000 in the item of Cyber Security with some of the other items.

- The agility of companies: Companies are agile.

When focusing only on the evaluation item "Cyber Security", the so-called five-eyes country rankings are 12th (Canada), 22nd (UK), 27th (USA), 31st (Australia), and 52nd (New Zealand). Those of other high GDP countries, China 10th, Japan 43rd, Germany 21st, India 33rd, France 34th, and so on. This result suggests the difficulty of cybersecurity measures taken by individual organizations in large countries. Moreover, the "Government cyber security capability" does not always support general security measures. For example, the ranking value of the Japanese government's cyber security capability is 24th.

As we see above, information security, including cybersecurity, is critically needed for any type of organization, and here we insist that establishing an information security policy plays an important role in ISMS (Information Security Management System) coping. However, it requires a lot of work and resources, both in finance and human, for its full establishment. In this chapter, after giving some issues on the information security policy, we introduce an automatically generated system that might benefit especially SMEs.

The rest of the chapter is organized as follows; a review of information security policy is coming up in the next section. Some general issues on security management and policy are described, along with a review of some papers on cybersecurity applying ontology, in the following section. The outline of the proposed system, including a brief notice on ontology, is explained as the methodology in the following section. The last section is on the conclusion and future works.

2. Information security policy

In this section, a basic concept of the information security policy is described, the recent situation of its establishment, especially in Japan, is presented, and some literature on the topic is reviewed.

2.1 What is the information security policy

The information security policy is sometimes considered the comprehensive and integrated system for implementing ISMS, where various types of controls and measures are incorporated. In an issue titled "Information security policy sample" published in 2016 by the Japan Network Security Association (JNSA), a layers document construction model is adopted as shown in **Figure 1** with content descriptions [8].

- Basic policy: A document that broadly declares to the public the stance on information security.

- Information security policy: A document that describes the policy for information security management. Clarify the structure, roles, and responsibilities for addressing information security.

- Information security measures regulations: Clarify the information security measures that should be introduced and followed, including daily operations.

- Information security measures procedure manual: Clarify the specific actions that should be taken on a daily basis using products that implement information security measures.

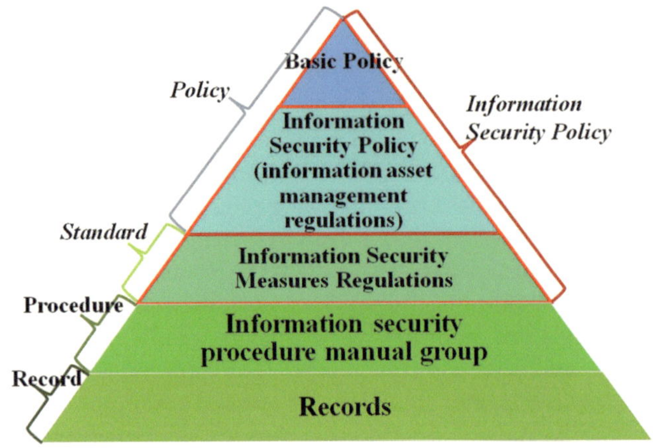

Figure 1.
5 Layer Model for Information Security Policy related Documents (Source: Information security policy sample, JNSA).

- Record: Records created in conjunction with compliance with information security measures and operational processes.

Leron Zinatullin shows a four-layer model consisting of "Policy", "Standard", "Guideline", and "Procedure", where "(Basic) Policy" is defined as a document providing a high-level overview of how organizational processes should operate in a secure manner [9]. He also described "Standard" as a regulation for the approach to security in the designated scope by preventing them from implementing conflicting or redundant solutions, and "Procedure" as a set of basic steps aiding the implementation of policies and standards.

In general, an information security policy describes the basic concept of what kind of information assets are to be protected and how to protect them from what threats. In other words, it is a systematization of the ideas and policies applied within the organization when building information security measures.

Here is a simpler model of three layers, with "Basic Policy", "Standard", and "Procedure" shown in **Figure 2**.

The short descriptions of the three-layer model are as follows;

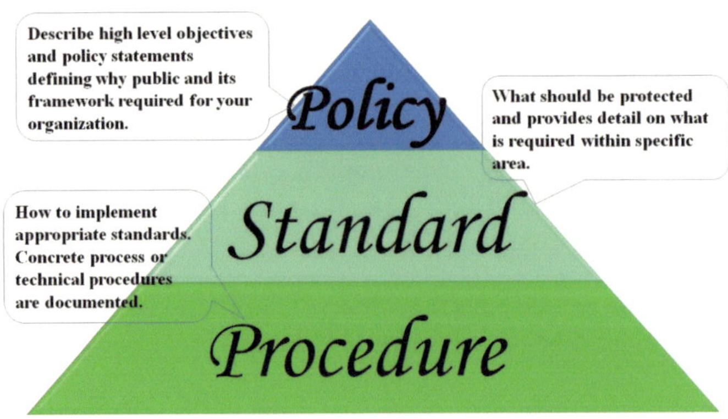

Figure 2.
3 Layer Model for Information Security Policy.

- Basic Policy: The document expresses the basic concept of information security measures, including declarations by representatives of organizations such as "Why information security is necessary?", "What kind of policy is used for information security?", and "What kind of policy will be used to handle customer information?".

- Standard: Specific measures and standards that are common to the entire organization will be cleared. In many cases, the standard provides general provisions for what kind of measures are to be taken, but in order to do so, it is necessary to analyze the information security risks in the organization.

- Procedure: The document (regulations, manuals, procedures, etc.) describes how to implement specific information security measures in each field and describes the details of the information security measures to be implemented for each countermeasure standard as a specific procedure.

2.1.1 Work flow for policy establishment

According to the 3 Layer model in **Figure 2** above, here describes the flow for establishing the information security policy along with **Figure 3**.

1. Establish the organization and the system: The involvement and responsibilities of the executives of the organization. A committee consisting of the heads of relevant departments, information system administrators, and a group of people with specialized knowledge of information security (e.g., the "Information Security Committee") shall be established, and its purpose, authority, and name shall be clarified. Almost all departments are listed, including information systems (LAN management sections, etc.), technical departments (experts with internal and external technical knowledge, etc.), auditing, human resources, accounting, public relations, and administration sections.

2. Formulation of the basic policy document: Declares that "the basic policy shall take measures to ensure the information security required for information systems", and the purpose, scope, and basic approach to information security measures. The basic policy defines the terms necessary to understand the policy, and since it determines the basic direction of information security, it should be noted that it will not be frequently updated.

3. Risk analysis: The risk analysis is to identify the information assets to be protected and assess the risks to them. Evaluate their importance from the CIA's point of view. Next, for all information assets, the frequency of occurrence of physical threats such as intrusion, destruction, failure, power outage, disaster, etc., unauthorized access, eavesdropping, computer viruses, technical threats such as falsification/erasure, DoS attacks, spoofing, and human threats such as mismanagement, elegance, fraudulent acts, improper management of passwords, etc. The magnitude of the damage when it occurs is examined, and the risk assessment of the information asset is used. At last, the measures against the risk are considered to be a method of reducing the magnitude of the damage.

4. Establishment of countermeasure standards: Individual countermeasures for each information asset are clarified through risk analysis, and their standards are determined by organizing and relocating them according to the following:

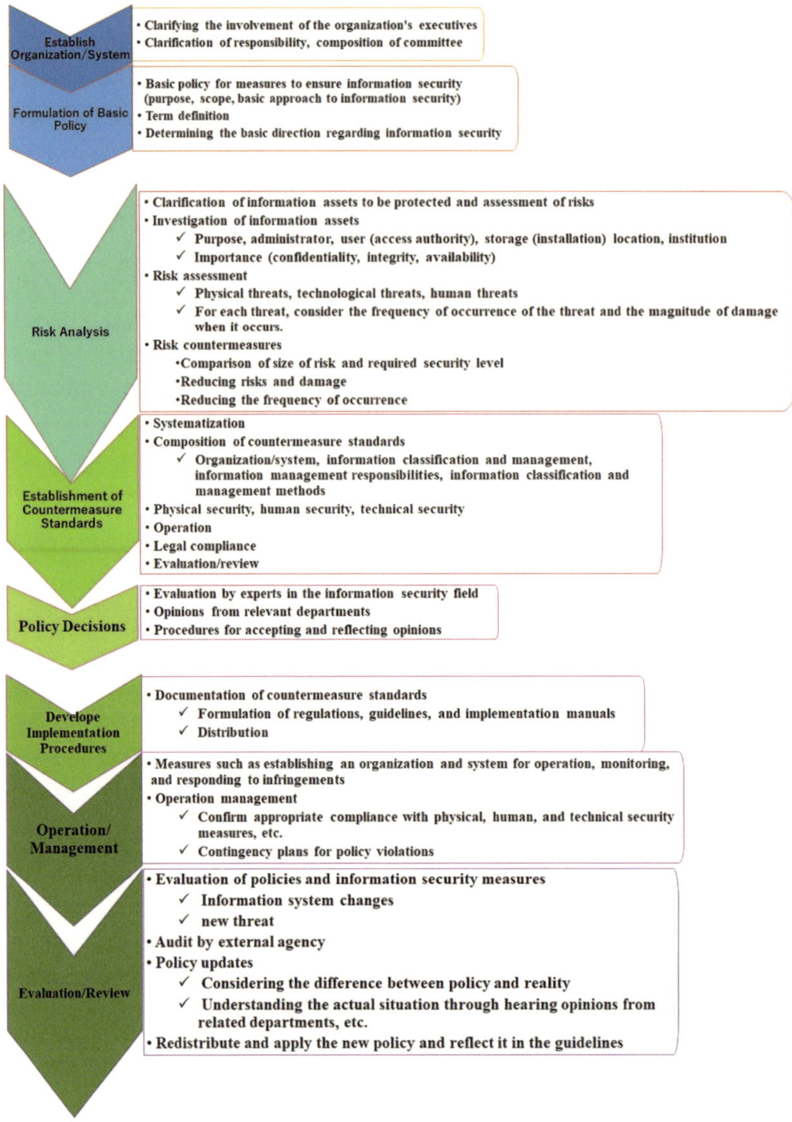

Figure 3.
Work Flow for Information Security Policy Establishment.

- Organization and structure
- Responsibilities and privileges for information, classification, and management of information
- Physical security
- Personal security: Roles, responsibilities and disclaimers, education and training, reporting of incidents and defects, password management, employment and contracting of part-time and temporary employees

- Technical security: computer and network management, access control, system development, installation, maintenance, etc., computer antivirus, collection of security information
- Operational security: Monitoring of information systems and confirmation of compliance with policies (operation management), points to note in operation management, countermeasures in the event of a breach, operation contracts by outsourcing
- Compliance with laws and regulations
- Response to information security violations
- Evaluation and review

5. Policy decision: Now that the basic policy and countermeasure standards have been established, consider it as an information security policy and make a decision on it. However, at the time of the draft policy, ask for evaluations by experts in the field of information security and opinions from related departments, etc. Based on the results, it is also necessary to establish procedures for accepting and reflecting opinions on the policy from the relevant departments at the operational stage.

6. Develop the implementation procedures: Now that the countermeasure standards in accordance with the basic policy have been decided, compile them into a document, formulate regulations, guidelines, and implementation manuals that can be referred to by each department, and distribute them.

7. Operation management: In order to ensure the operational management of information security policy, it is necessary to appropriately take measures such as establishing an organization and system, monitoring, and responding to breaches. If a policy violation that may cause serious problems is reported, it will be handled according to the contingency plan, conduct drills, review the precautions to be taken in communication, investigation, and response in the event of a breach, and develop a plan to prevent a recurrence.

8. Evaluation and review: Evaluate and review the countermeasure standards on a regular basis based on changes in the information system and the emergence of new threats. An audit by an external organization is also necessary, but in that case, the weakness of the information system will be known to the relevant organization, so carefully consider the reliability and select an audit organization. When updating the policy, it may be necessary to fully consider the differences between the policy and the actual situation, to grasp the actual situation by listening to the opinions of the relevant departments, etc., and to conduct a new risk analysis.

2.1.2 Basic policy document

As an example of items in the basic policy document, first have a look at them in "The Importance of Information Security in Financial Institutions and

Countermeasures" [10]. Though this report is very old, items picked up still have universality for the basic policy document, for example.

Purpose and Scope of Information Security Measures: Basic concept of information security measures, scope (basically for all organizations) and reason why information and systems to be protected, identification of information assets to be protected and people who handle them, and priority standard evaluating information assets and systems to be protected.

- Promotion of Information Security Measures: Involvement and responsibilities of management officers, the appointment of an officer in charge of information security and establishment of a security department, checks by the legal department regarding laws and regulations, compliance, use of external consulting, etc.

- Operation of Information Security Measures: Expected information security risks and their management, the decision-making process for the implementation of information security measures, procedures for reviewing information security measures, and outline of specific information security measures.

- User management and information security education: Responsibilities of each officer and employee and arrangements in the event of a violation (penalties, etc.), checking the status of compliance with information security measures, and promoting awareness of information security policy.

- Response to failures in the crisis management system.

- Other: procedures for periodic review of the information security policy.

In the "Guidelines for Information Security Policies in Local Governments" formulated by the Ministry of Internal Affairs and Communications on March 30, 2001, and revised on September 25, 2018, [11], the following example sentences are written as a basic policy item.

1. Purpose: The purpose of this basic policy is to establish basic matters regarding the information security measures implemented in order to maintain the CIA of information assets held by our organization.

2. Definitions of words, terminologies, etc.

3. The following threats are assumed as threats to the target information assets, and proper information security measures should be implemented.

    - Leakage, destruction, falsification, or deletion of information assets due to intentional factors such as unauthorized access, virus attacks, denial of service attacks, cyber-attacks, intrusion by outsiders, theft of important information, internal fraud, etc.

    - Unauthorized removal of information assets, violation of regulations such as the use of unauthorized software, inadequacies in design and development, program defects, operation and configuration errors, maintenance deficiencies, inadequacies in internal and external audit functions, leakage, destruction, deletion, etc. of information assets due to unintentional factors such as inadequate outsourcing management, management defects, equipment failures, etc.

- Earthquakes, lightning strikes, suspension of services and operations due to disasters such as fires, etc.
- Dysfunction of system operation due to shortage of personnel due to large-scale and widespread diseases.
- Spillover effects from infrastructure failures such as interruptions in power supply, communication interruptions, and interruptions in water supply.

4. Scope of application

- Scope of administrative agencies: The administrative agencies to which this basic policy applies are internal departments, administrative committees, parliamentary secretariats, fire departments, and local public enterprises.
- Scope of information assets: The information assets covered by this information security policy are as follows; networks, information systems and related equipment, electromagnetic recording media, information handled by networks and information systems (including printed documents), and system-related documents such as information system specifications and network diagrams.

5. Employees, part-time employees, and temporary employees must have a common understanding of the importance of information security. And comply with the information security policy and information security implementation procedures in the performance of their duties.

6. Information Security Measures: The following information security measures are taken to protect information assets from the threats described above in 3.

- Organizational structure: Establish an agency-wide organizational structure to promote information security measures for the information assets.
- Classification and management of information assets: The information assets held by the organization shall be classified according to the CIA, and information security measures shall be implemented based on the classification.
- Improvement of the resilience of the information system as a whole: The following three measures will be taken for the information system as a whole. (1) Prevent leakage of stakeholders' information in the system, (2) implement detoxified communication in intranet or extranet connection systems, and (3) implement information security measures in Internet connection systems, cloud computing systems, etc.
- Physical security: Take physical measures for the management of communication lines and personal computers of employees.

- Human security: Establish matters to be complied with by employees, etc., regarding information security, and take personnel measures such as providing sufficient education and enlightenment.

- Technical security: Take technical measures such as computer management, access control, anti-malware measures, and countermeasures against unauthorized access.

- Operation: Take measures for the operation of the information security policy, such as monitoring the information system, checking the status of compliance with the information security policy, and ensuring security when outsourcing.

- Use of external services: In the case of outsourcing, select an outsource, conclude a contract that specifies information security requirements, and confirm that the necessary security measures are ensured by the outsource. In the case of using external services in accordance with the terms and conditions, establish regulations and take measures for the use of external services. When using social media services, establish operating procedures for social media services, stipulate the information that can be transmitted on social media services, and determine the person responsible for each social media service to be used.

- Evaluation and review: In order to verify the status of compliance with the information security policy, conduct information security audits and self-inspections, and improve operations to enhance information security.

7. Implementation of information security audits and self-inspections: Information security audits and self-inspections are carried out periodically or as necessary to verify the status of compliance with the information security policy.

8. Review of the information security policy: If it becomes necessary to review the information security policy as a result of the information security audit and self-inspection, or if new measures are required to respond to changes in the information security situation, the information security policy will be reviewed.

9. Formulation of information security standards: In order to implement the measures stipulated in 6, 7, and 8 above, formulate information security measures standards that stipulate specific compliance matters and judgment criteria.

10. Formulation of information security implementation procedures: Information Security Implementation Procedures shall be formulated based on the Information Security Measures Standards, which stipulate specific procedures for implementing information security measures.

For universities and other educational and research institutions, the Ministry of Education, Culture, Sports, Science and Technology of Japan (MEXT) obliges national universities to prepare one. According to the "Summary of the Academic Information Infrastructure Fact-finding Survey" conducted by the MEXT in 2019 [12], security policies are formulated by all universities in 86 national universities and 86 (92.5%) in 93 public universities. Talking about private universities, 73.4% of 613 universities formulated a kind of security policy statement at that time.

Although information security policies are required for universities and other educational institutions, their basic policy documents published on their websites are much simpler than those described above.

3. Some issues on information security management and policy

ISO/IEC 27000 family [13], some of which are based on BS7799, is one of the well-known ISMS frameworks. In 3.1.24 of the latest version of ISO/IEC 27002: 2022, quoting ISO/IEC 27000:2018, 3.53, the policy is determined as "intentions and direction of an organization, as formally expressed by its top management". Since ISO/IEC 27002 is the guideline for organizational information security standards and management by giving a code of practice for information security controls, policy is handled as one of the controls in parallel with some others such as asset classification, personal security, physical and environmental security, etc.

According to the model in **Figure 2**, a supporting system for generating the security policy is proposed, then an initial program was created as a prototype [14]. The flow of the system is depicted in **Figure 4**.

- Explanation of information security policy itself in several media such as text, audio, video, etc., then its necessity is explained to those who are responsible for the organization.

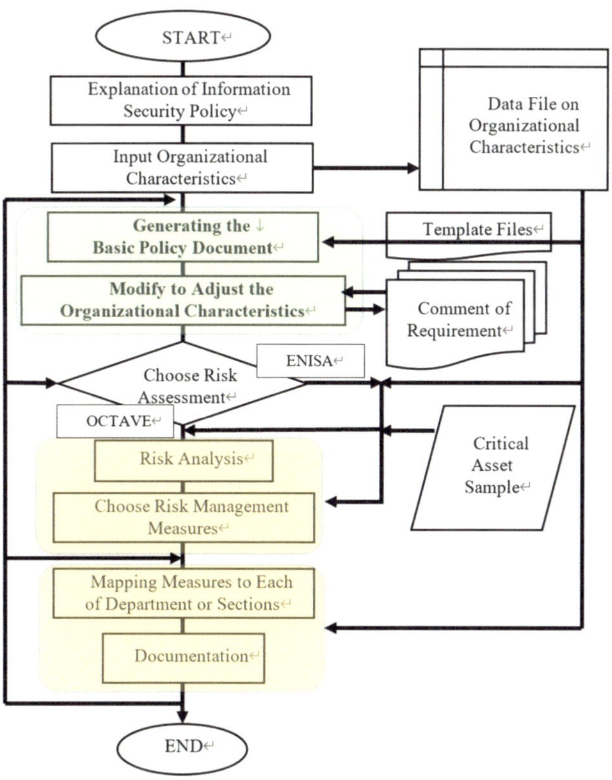

Figure 4.
Overall Flow of Information Security Policy Establishment (Source: Nagata [14]).

- Input the organizational characteristics.
- Generating basic policy using a template consisting of several items by referring to organizational data file and displaying with some comment on requirements when adopting this expression or word.
- In case of going to the standard stage, choose one of the risk analysis methods from OCTAVE [15], or ENISA's system [16]. They output a set of mitigation controls. A precise method to select effective risk mitigation measures is proposed in [17], and that for identifying information-related assets is in [18].
- Summarizing and documenting the procedures for each department, then the document will be completed after hearing the opinions of each department.
- The arrows from down to up represent the PDCA cycle of overall security policy.

In order to reflect organization's characteristics, some systems with ontology-based processes would be more effective.

3.1 Review of some ontology-based information security management systems

Gruber [19], noted that.

> *"Ontology is an explicit specification of a conceptualization. The term is borrowed from philosophy, where an ontology is a systematic account of existence. For knowledge-based systems, what 'exists' is exactly that which can be represented."*

He also claimed that sharing a common understanding of the structure of information among people or software agents is one of the more common goals in developing ontologies.

In the guide for ontology development published by Noy and McGuiness [20], an ontology is denoted as a formal explicit description of concepts in a domain of discourse, properties of each concept describing various features and attributes of the concept, and restrictions on slots.

Herzog et al. [21] gave a security ontology built upon classical components of risk analysis and their relations to each other.

Since, in an organization, information assets play a central role in establishing an information security policy, creating a detailed information asset ontology becomes important. Zeb et al. [22], proposed an ontology-supported Asset Information Integrator System (AIIS) to help industry experts exchange their tangible capital assets information and transform the way they were exchanged at that time between the municipal and Canadian provincial governments. They presented the ontology development methodology in ten steps as the hybrid version of several former works.

Aiming to assist in determining inherent attributes of IT assets that can assist in the process of IT assets risk value assessment, Adesemowo et al. [23], published a paper on IT assets ontology, where assets are divided into "Personnel", "Network", "Services", "Data", "Hardware", "Software", and "Information".

Fenz et al. proposed an ontological mapping of ISO/IEC 27001 [24], and ISO/IEC 27002 [25], as an ontology-based ISMS policy implementation. The overview of creating security ontology based on ISO/IEC 27002 is described in **Figure 5**.

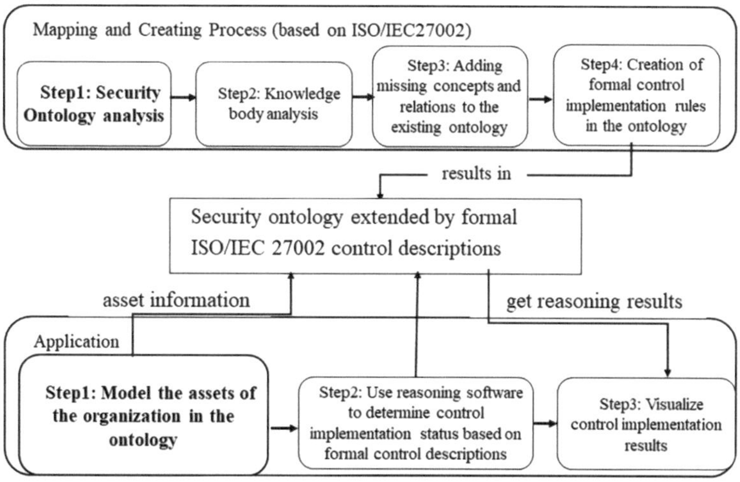

Figure 5.
Overview of Mapping ISO 27002 in the Ontological Structure and Applying the Results (Source: Fenz et al. [25]).

Pereira and Santos [26], represented a somehow different conceptual framework for information security ontology from that of Fenz et al. [25]. Although they have "Asset", "Threat", "Vulnerability", and "Control" as common classes, relations between the two of them are distinct. For instance, "Control protect asset" in Pereira and Santos, whereas "Control (is) implemented (in) Asset" in Fenz et al. Thus, the configuration of the ontology will vary depending on the adopted criteria, organizational characteristics, purpose, way of thinking, and so on.

Almost all the policy-based ontologies aim to present mitigation measures against risks and threats and means to compensate for vulnerabilities. For that purpose, it is necessary to create a detailed ontology coping with the characteristics of the organization. Unified Process for Ontology (UPON) was proposed by Nicola et al. [27], for building a large-scale ontology in four workflows as "Requirements", "Analysis", "Design", "Implementation", and "Test" by a domain expert and knowledge expert. Methods, UPON, may be helpful for creating a precise ontology.

A pioneering policy management framework named KAoS by Uszok et al. [28], uses semantically rich ontological representation and reasoning composed of three layers, "Human Interface Layer", "Policy Management Layer", "Policy Monitoring and Enforcement Layer". The basic form of KAoS policy is as follows:

[Actor] is [constrained] to perform [controlled action] which has [any attributes].

Tonti et al. [29], discussed the implementation of the enforcement system of OWL policies using the KAoS policy framework into multi-agent systems built on top of the JDK1.4.

4. Generation system of information security policy with ontology

As we see in Section 3, an automated basic information security policy statement generation system described in **Figure 4** is proposed for embedding into the existing Java application program.

Although some systems incorporating ontology for information security enforcement are proposed and several methodologies mentioned in the previous section are useful and effective for organizational information security establishments, those are concerned with the creation of ontologies that reflect policies and the methodology of automatically configuring means to ensure information security using ontologies.

Here we mention the additional proposed phase of the system in the upper part of **Figure 4** for creating a general basic statement by applying organization-related ontology [30]. **Figure 6** depicts the outline of the phase in the upper part of which the system queries corresponding ontology by using input organizational essential data, then constructs a set of candidate phrases of basic policy. Policymakers try to adjust or modify the represented policy with the help of the ontology again in the lower part.

The key point is to apply ontology, a method of creating competency questions from sample sentences, and configure an ontology that can respond to them is also proposed.

Organizational ontology also plays an important role in this phase, but what is needed here is the creation of basic policy phrases, not the detailed ontology that is treated in many studies. To implement the phase, ontologies for different types of organizations must be created in advance.

Here describe the method for ontology creation.

Step 1 Gather sample phrases of basic policy and then classify the set of them into each of the typical items in the former implemented system, such as "Concept and Purpose", "Scope of Application", "Definition of terms", "Composition/Positioning", "Management system", "Role/Responsibility", and "Basic requirements".

Step 2 Analyze the sample policies to get competency questions (CQ) for ontology. For example, if there is a sample phase reading.

"The CEO serves as chairman of the information protection committee and is responsible for information security within the organization", then CQs will be like as follows:

CQ1: "Is there a body for ensuring information security?"

CQ2: "Who serves as chairman of the ISMS committee?"

CQ3: "Is the chairman ultimately responsible for ISMS?"

Step 3 Configure each of the ontologies according to the type of organization. These types of organizations are pre-determined relatively broadly according to business conditions, such as universities, high schools, research institutes, manufacturing

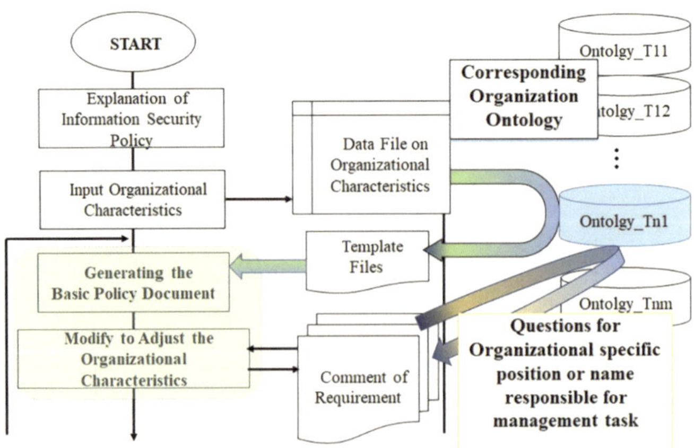

Figure 6.
Improved Version of the First Stage of Former System (Source: Nagata [30]).

industries, distribution industries, etc., as well as their management style and scale. Then create an ontology that will be common to each of these types.

In step 3, we can apply existing ontologies for general matters such as FOAF ontology for academic organizations [31].

5. Conclusions and future works

This chapter deals with information security as a risk to organizations. First, the current state of information risks such as cyber crimes and their effects are described, and an overview of information security policies as a means of ensuring information security for organizations.

In the second half, we looked at specific information security policy contents and an automatic generation system using ontology.

Some of the proposed automated basic information security policy statement generation systems are embedded into existing Java application programs. However, ontology configuration is a time-consuming and skill-intensive process, and the validity assessment of the prepared statement will be necessary.

Although application programming by Java is still in the development stage, we think that the direction for proceeding to the countermeasure standard creation stage following this basic policy stage is indicated.

Abbreviations

AIIS	Asset Information Integrator System
BEC	business email compromise
BS	British Standard
CEO	Chief Executive Officer
CIA	Confidentiality, Integrity, and Availability
CQ	competency questions
$	U.S. dollar
DoS	denial of service
ENISA	European Network and Information Security Agency
FBI	Federal Bureau of Investigation
FOAF	friend of a friend
GDP	gross domestic product
GRPS	Global Risks Perception Survey
IC3	Internet Crime Complaint Center
IMD	International Institute for Management Development
ISMS	information security management system
ISO/IEC	International Organization for Standardization/International Electrotechnical Commission
JDK	Java development kit
JNSA	Japan network security association
KAoS	knowledge acquisition in automated specification
LAN	local area network
MDR	Managed detection and response
MEXT	The Ministry of Education, Culture, Sports, Science and Technology of Japan

OCTAVE	operationally critical threat, asset, and vulnerability evaluation system
OWL	web ontology language
PDCA	Plan, Do, Check, and Action
SMEs	Small or medium-sized enterprises
UPON	Unified process for ontology

Author details

Kiyoshi Nagata
Faculty of Business Administration, Department of Business Management, Daito Bunka University, Tokyo, Japan

*Address all correspondence to: nagata@ic.daito.ac.jp

References

[1] The Global Risks Report 2023, 18th Edition, World Economic Forum. Available from: https://www.weforum.org/reports/globalrisks-report-2023/ [Accessed: January 24, 2024]

[2] The Global Risks Report 2022, 17th Edition, World Economic Forum. Available from: https://www3.weforum.org/docs/WEF_The_Global_Risks_Report_2022.pdf, [Accessed: January 24, 2024]

[3] Cybercrime Expected To Skyrocket in Coming Years, Anna Fleck. Available from: https://www.statista.com/chart/28878/expected-cost-of-cybercrime-until-2027/ [Accessed: January 24, 2024]

[4] 2022 Official Cybercrime Report. Available from: https://www.esentire.com/resources/library/2022-official-cybercrime-report [Accessed: January 24, 2024]

[5] FBI Internet Crime Report 2022 by Internet Crime Complaint Center. Available from: https://www.ic3.gov/Media/PDF/AnnualReport/2022_IC3Report.pdf [Accessed: January 24, 2024]

[6] IMD World Digital Competitiveness Ranking 2023. IMD World Competitiveness Center. Available from: https://www.imd.org/centers/world-competitiveness-center/rankings/world-digital-competitiveness/ [Accessed: January 24, 2024]

[7] IMD World Digital Competitiveness Ranking 2022, IMD World Competitiveness Center. Available from: https://www.imd.org/wp-content/uploads/2023/03/digital-ranking-2022.pdf [Accessed: January 24, 2024]

[8] Information security policy sample, Japan Network Security Association. Available from: https://www.jnsa.org/result/2016/policy/data/policy_gaiyou.pdf, [Accessed: January 2024]

[9] Zinatullin L. The Psychology of Information Security. IT Governance Publishing; 2016. ISBN-13: 978-1849287890

[10] Japan Bank Survey Monthly Report, Bank of Japan, April 18, 2000. Available from: https://www3.boj.or.jp/josa/past_release/chosa200005c.pdf, [Accessed: January 24, 2024]

[11] Guidelines for Information Security Policies in Local Governments. Available from: https://www.soumu.go.jp/main_content/000575052.pdf [Accessed: January 24, 2024]

[12] Summary of the Academic Information Infrastructure Fact-finding Survey, Ministry of Education, Culture, Sports, Science and Technology of Japan. Available from: https://www.mext.go.jp/content/20200721-mxtjyohoka01-000005760_2.pdf [Accessed: January 24, 2024]

[13] ISO/IEC 27000 family-Information security management, ISO. Available from: https://www.iso.org/standard/iso-iec-27000-family, [Accessed: January 24, 2024]

[14] Nagata K, Kigawa Y. Construction of support system for information security policy. In: Proceedings of the 20th Asia Pacific Industrial Engineering and Management Systems Conference 2019, Kanazawa, Japan, December 02-05, APIEMS. 2019. pp. 942-947

[15] Alberts C, Dorofee S, Stevens J, Woody C. In OCTAVE-S Implementation Guide, Version 1.0, CMU/

SEI-2003-HB-003. Carnegie Mellon Software Engineering Institute; 2005. Available from: https://apps.dtic.mil/sti/pdfs/ADA430801.pdf [Accessed: January 24, 2024]

[16] Information Package for SMEs- With examples of Risk Assessment/ Risk Management for two SMEs-, George Patsis. Available from: https://www.enisa.europa.eu/publications/information-package-for-smes/ [Accessed: January 24, 2024]

[17] Nagata K, Kigawa Y, Cui D, Amagasa M. Method to select effective risk mitigation controls using fuzzy outranking. In: Proceedings of the 9th International Conference on Intelligent Systems Design and Applications, Pisa, Italy, November 30–December 02, 2009. IEEE Computer Society; 2009. pp. 479-484. ISBN: 978-0-7695-3872-3. Available from: https://dl.acm.org/doi/10.1109/ISDA.2009.186 [Accessed: January 24, 2024]

[18] Nagata K. Chapter 6, construction of effective database system for information risk mitigation, security enhanced application for information systems. In: Security Enhanced Applications for Information Systems. London, UK: INTECH Open Access Publisher; 2012. pp. 111-130. ISBN: 978-953-51-0643-2. Available from: https://www.intechopen.com/chapters/37311 [Accessed: January 24, 2024]

[19] Gruber TR. A translation approach to portable ontology specifications. Knowledge Acquisition. 1993;**5**(2):199-220. DOI: 10.1006/knac.1993.1008

[20] Noy NF, McGuiness DL. Ontology Development 101: A Guide to Creating Your First Ontology. Stanford Knowledge Systems Laboratory Technical Report KSL-01-05; 2001. Available from: https://protege.stanford.edu/publications/ontology_development/ontology101.pdf [Accessed: January 24, 2024]

[21] Herzog A, Shahmehri N, Duma C. An ontology of information security. Internationl Journal of Information Security and Privacy. 2007;**1**(4):1-23

[22] Zeb J, Froese T, Vanier D. An ontology-supported asset information integrator system in infrastructure management. Built Environment Project and Asset Management. 2015;**5**(4):380-397

[23] Adesemowo AK, Solms R, Botha RA. ITAOFIR: IT asset ontology for information risk in knowledge economy and beyond. In: Proceedings of 11th International Conference, Global Security, Safety and Sustainability: The Security Challenges of the Connected World 2017, January 18-20. London, UK: Springer; 2017. pp. 173-187

[24] Fenz S, Goluch G, Ekelhart A, Riedl B, Weippl E. Information security fortification by ontological mapping of the ISO/IEC 27001 standard. In: 13th Pacific Rim International Symposium on Dependable Computing (PRDC 2007). Melbourne, VIC, Australia: IEEE; 2007. pp. 381-388. DOI: 10.1109/PRDC.2007.29

[25] Fenz S, Plieschnegger S, Hobel H. Mapping information security standard ISO/IEC 27002 to an ontology structure. Information & Computer Security. 2015;**24**(5):452-473

[26] Pereira T, Santos H. An ontology approach to information security management. In: Sartori F, Sicilia MÁ, Manouselis N, editors. Metadata and Semantic Research. MTSR 2009. Communications in Computer and Information Science. Vol. 46. Berlin, Heidelberg: Springer; 2012. pp. 368-375. DOI: 10.1007/978-3-642-04590-5_17

[27] Nicola AD, Missikoff M, Navigli R. A software engineering approach to ontology building. Information Systems. 2009;**34**:258-275

[28] Uszok A, Bradshaw JM, Jeffers R. KAoS: A Policy and Domain Services Framework for Grid Computing and Semantic Web Services. In: Trust Management. iTrust 2004. Lecture Notes in Computer Science. Vol. 2995. Berlin, Heidelberg: Springer; 2004. pp. 16-26

[29] Tonti G, Montanari R, Bradshaw JM, Bunch L, Jeffers R, Suri N, et al. Automated generation of enforcement mechanisms for semantically-rich security policies in java-based multi-agent systems. In: IEEE First Symposium on Multi-Agent Security and Survivability, 2004. Drexel, PA, USA: IEEE; 2004. pp. 11-20. DOI: 10.1109/MASSUR.2004.1368413

[30] Nagata K. Automatic generating system of information security policy. Athens Journal of Technology and Engineering. 2023;**10**(4):227-236

[31] Kalem E, Martiri E. FOAF-academic ontology: A vocabulary for the academic community. In: Proceedings 2011 Third International Conference on Intelligent Networking and Collaborative Systems. Fukuoka, Japan: IEEE; 2011. pp. 440-445. DOI: 10.1109/INCoS.2011.94

Chapter 3

Risk Assessment and Machine Learning Models

Naimeh Borjalilu

Abstract

Safety management system for operational domains with extensive scope is a challenging issue. In the current safety management system literature, there are not efficient methods to implement subjective inferences of organization experts to analyze data and predict the performance of safety and quality management systems by using all available data and identify weakness points to improve organizational safety and quality process. Machine learning is a subset of artificial intelligence that involves the algorithms and model development to allow safety and quality management systems that improve their performance over time automatically. The benefits of machine learning include efficiency improvement, reduced costs, improved decision-making, and increased innovation to study the future of risk management. The existence of massive data in operational sectors is a critical challenge for machine learning implementation in safety risk management systems. A safety data pool from all occurrences and hazards should be designed to improve and know about the future of safety management system, and then design the system to assess, analyze, verify, and predict the safety assessment result to decide proper management system decisions. In this study, the machine learning method implementations are proposed for risk management in operational domains.

Keywords: risk assessment, machine learning, prediction, safety management, system, management

1. Introduction

Serious incidents or accidents often occur during flight operations, which have severe consequences. The International Air Transport Association (IATA) has developed standards for analyzing safety data to monitor and emphasize the scope of flight safety concerns. There are some areas that benefit from this study in reviewing and analyzing the safety index, such as authorities and insurance companies.

ICAO itself recognized the framework for the safety management system (SMS), **Figure 1** show the components and elements of SMS framework.

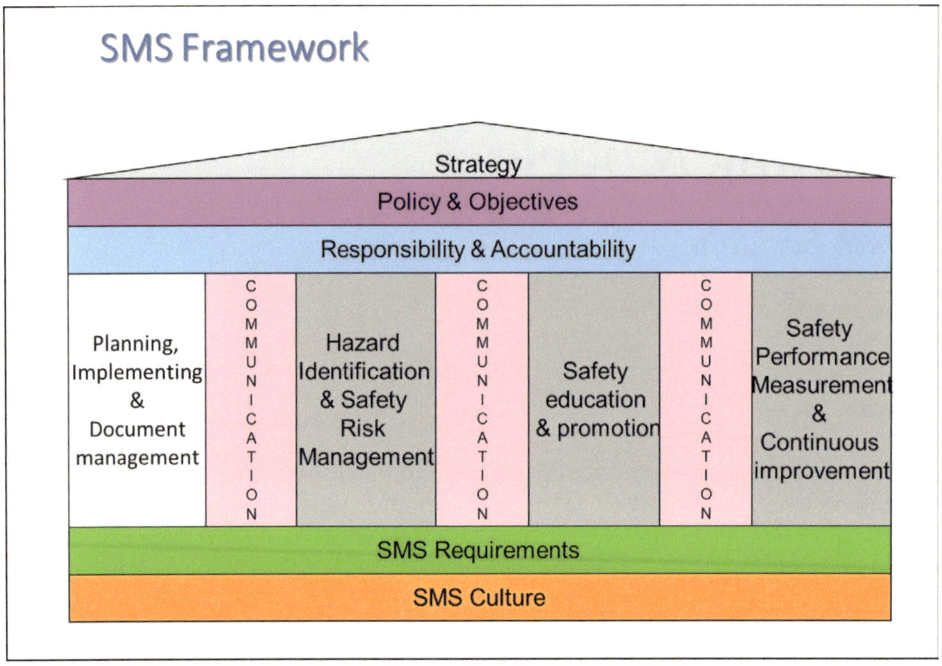

Figure 1.
Safety management system component.

1. Safety policy and objectives.

2. Safety risk management.

3. Safety assurance.

4. Safety promotion.

Safety is clarified as "the state in which the possibility of harm to persons or of property damage is reduced to, and maintained at or below, an acceptable level through a continuing process of hazard identification and safety risk management." SMS implementation starts with hazard identification. Hazards and their related outcome are essential to the implementation of safety and quality risk management [1]. A safety risk is measured by the likelihood and severity appointed to the safety consequence. Safety risk mitigations are proposed to decrease the level of safety risk. There are a few possible risks that can occur during a production process [2].

The implementation of the safety risk management process needs to identify hazards. Obvious hazard identification and their related outcome are necessary to implement safety risk management. A hazard is generically identified by safety experts as a state or a statute with the potential to bring death, injuries to personnel, harm to equipment or structures, loss of material, or reduction of the ability to carry out activities. Hazards are available at all stages in the organization and will be identified or detected by using reporting systems, inspections, or audits. As a result, hazards should be detected before they cause accidents, incidents, or other safety-related occurrences. A main method to identify proactive hazards is a voluntary hazard/incident reporting system.

Safety risk management includes the evaluation and mitigation of safety risks. The goals of safety risk management are risk assessment, which is related to hazard identification, and effective and suitable mitigations that need to be implemented. The key component of the safety management process is safety risk management at both the regulatory body and product/service provider level.

In recent years, machine learning algorithms have solved domain-specific problems in various fields. Analyze and predict safety risk assessment in the aviation industry by machine learning algorithms usage is increased and also used.

For decision-making, business understanding, data understanding, data preparation, modeling, evaluation, and deployment are the methods to include machine learning [3]. Risk assessment usage is applied in many application scopes, and many frameworks, methods, and particular applications have been studied in the scientific literature [4].

One of the most important methods to provide significant benefits for businesses and individuals is machine learning, which has been a rapidly growing field in recent years. From improved efficiency to increased innovation, machine learning has the potential to revolutionize the way to increased innovation. Lack of understanding and expertise, cost of implementation, and ethical concerns when adopting machine learning are challenges and issues to use machine learning; despite these challenges, the potential benefits of machine learning are too great. This chapter presents the importance of machine learning in implementing a safety management system used to predict the safety risk trend in the aviation industry. It also initiates the appeal that is utilized by the authors of this book to arrange the chosen literature into different classifications.

2. Safety risk assessment model

Safety risk assessment is a key point in safety-related industries. However, it encounters a number of concerns, partially associated with technological progress and necessary increment. There is currently a demand to assess risk, improve past lessons learning, and identify methods to analyze relevant data. Data are collected with sufficient ability to converse with unexpected events and supply suitable support to empower safety risk management [5].

The likelihood and severity of the consequence or outcome from an identified hazard are measured in safety risk assessment. It is necessary to provide and propose risk mitigation for the identification of such layered consequences.

2.1 Safety risk probability

"The likelihood or frequency that a safety consequence or outcome might occur" is defined as safety risk probability. The probability assessment is the first step in the process of controlling safety risks.

2.2 Safety risk severity

"The harmness which is happened as a consequence or outcome of the detected hazard" is defined as safety risk severity. Once the probability assessment has been completed, the safety risk severity assessment is the next step is to consider the potential consequences related to the hazard.

2.3 Safety risk tolerability

The safety risk probability and severity assessment process can be applied to extract a safety risk index. The index made using the method identified above consists of an alphanumeric designator, indicating the results of the probability and severity analysis.

A safety risk assessment matrix is used to obtain a risk index and then exported to a safety risk tolerability matrix. Safety risk management comprises the assessment and risk analysis to suggest and implement mitigation of safety risks.

The assessment of the risks is the main target of safety risk management at both the state and product/service provider levels, and it is associated with identified hazards to develop and implement efficient and effective mitigations.

2.4 Safety management system

A systematic attitude to manage safety, including the necessary organizational structures, accountabilities, policies, and procedures.

2.5 Risk mitigation

The procedure of compounding defenses or preventive controls to lower the severity and/or likelihood of a hazard's projected outcome/consequence.

2.6 Human factors (HF)

The main and specific part of the safety analysis process are SSPs and SMSs with human and organizational factors, which are effective risk management systems. Human factors (HF) is an essential point to assure that existing or recommended defenses have been effective to decrease the safety risk. Where necessary, a supplementary HF analysis may be conducted to support that particular risk mitigation exercise/team. Analysis and error definition and error categorization use the human error model to allow the root cause of all hazards and better knowing the result. This conception ensures the sufficient fulfillment of a root cause analysis. The result of human factors analysis can help the safety expert identify root causes and make better decisions for top managers. A more comprehensive and in-depth mitigation process is the important human factors perspective results.

3. Machine learning implementation

The undertaking applications and prompt development in many scope of machine learning (ML) are the reasons which is received a lot of attention recently.

Machine learning (ML) is a method that can learn a task without being explicitly programmed. This makes it very strong and attractive. ML algorithms spontaneously learn an activity from data. They perform some form of inductive inference whereby, after being trained on a dataset, they make predictions for inputs outside the examples observed in the dataset. ML can achieve feasible and cost-effective tasks that could be cost-prohibitive to solve with explicit systems [6].

Machine learning (ML) has received a lot of attention recently due to its agile progress and promising applications in many scopes [7].

Data-driven analysis is an essential tool for both operational efficiency and improved safety and quality practices. Growth in the industry, which is connected with technology progress and data science techniques, has led the operational industry to a more data-driven approach. Safety and quality experts should access to many sources of safety and quality-related information to assess and analyze—significant methods currently available leverage large data sets to identify anomalous business behavior. In recent years, data mining and machine learning methods have been implemented to analyze safety, incident and accident investigation, and error detection in the community. There are many techniques in machine learning (**Figure 2**).

Figure 2 shows that there are two techniques in data mining models: conventional and data-driven techniques for smart grid analysis. Each method has a specific benefit according to the application scope.

Safety and quality data are collected and processed from many resources such as operational safety and quality assessment and assurance. Safety and quality reports reside in both the specific Safety Action Program and the National Systems coordinated Safety Reporting Systems initiative. This process contains data preprocessing, a high-frequency event evaluation, feature vector generation, algorithm implementation, and post-processing of the results [8].

Machine learning (ML) recognize to be one of the only technically and economically practical resolving to robotize some complicated activities usually discover by humans, such as driving vehicles and recognizing voice. However, these methods face new potential risks and have only been accomplished in systems where the benefits of the methods are known to be worth this increase of risk [9]. Some researchers categorize ML safety into three plans: (1) intrinsically safe methods, (2) method performance and strength, and incorporate (3) run-time error detection methodology [7].

Machine learning as a methodology can detect criteria and schemes in data, predict future data, and help experts make better decisions under uncertainty. Machine

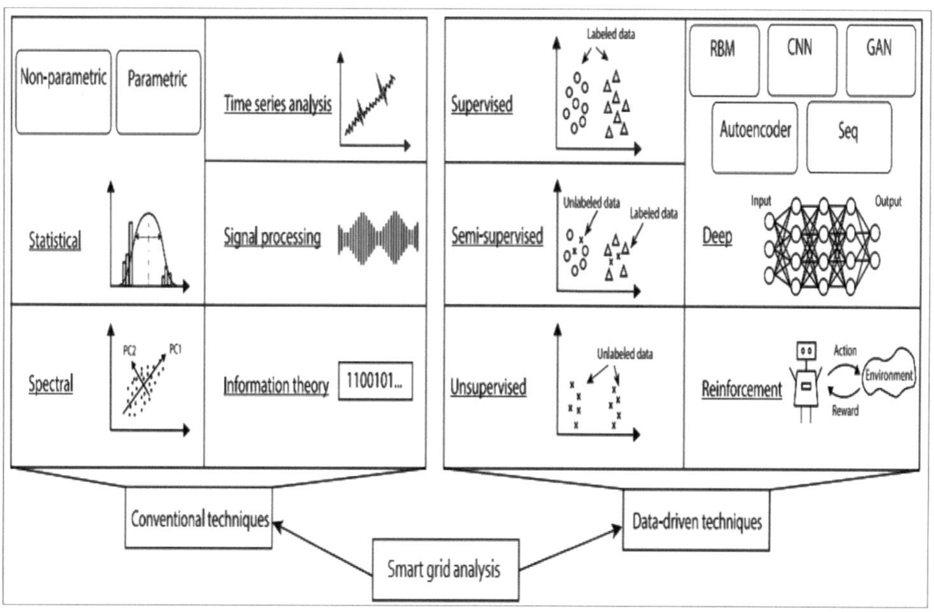

Figure 2.
Data mining methods.

learning use statistics sciences and computers to evaluate complicated functions and a decreased reliance on proving confidence intervals around these activities [10]. Machine learning is applied as suitable tool to achieve human-like choice or even super-human performance with automation on specific activities. Data-driven approach is currently have enormous challenge in all industry scope. Enormous growth show that machine learning (ML) has many advancements in computer vision and other fields. Training a neural network is done by using very big datasets to implement machine learning [11].

Currently, a variety of machine learning methods can be used to implement SMS phases including: risk identification, risk analysis, and risk evaluation. Machine learning use data (such as historical and real-time data) for providing inputs to traditional risk assessment techniques. Machine learning is used and applied in many industries such as automotive, aviation, construction, and railways. Paltrinieri et al. [5] studied the knowledge gap by performing a structured review of relevant literature on machine learning usage for risk assessment. The results show that 11 journal papers were published in the *Journal of Accident Analysis and Prevention*, which makes it the most contributing journal reference. The United States of America, Chinese, and South Korean institutes were the highest level of following affiliations. The adoption of machine learning to implement risk assessment is studied in the automotive industry (over 20% of published articles). The most popular machine learning algorithm selected to accomplish risk assessment was artificial neural networks (ANNs) which are used by support vector machines (SVMs). Historical datasets are practiced in more than 70% of papers, and real-time data to construct and design the machine learning model are implemented in more than 20% of paper. A case-study approach to machine learning model implementation is used in about half of the proposed methods, and about one-fourth have applied their models in a real-world setting.

The results of this review show that the machine learning techniques usage to assess the future of risk is a new approach to study and has an increasing trend in academic publications. Classic risk assessment is supported by machine learning methods and data-driven preparation as inputs. In the future, the operational domains in the industry will need to assess risk in a real-time manner with machine learning methods adoption. The Safety regulatory bodies can use machine learning in risk assessment to validate procedures [5]. Data mining and its key enabler, machine learning, and techniques are the main and usable tools for predicting because these algorithms effectively identify and capture complex patterns and relationships. However, they can account for variable-related assumptions and other limitations, such as a lack of ability to explain things and transparency. Some factors can be present in the data such as environmental factors, human performance, and instrument.

To predict risk categories by using safety and quality data, machine learning algorithms were implemented to classify records in the data set. The prediction of risk categories is determined by using multiple models to be implemented and tested.

To start, features were selected for inclusion in the model. Once features were selected, data was restructured into a valid template for each model's feature. Each model applied by machine learning shall be trained, and tested, and has its own strengths and weaknesses. Overall, when selecting a suitable model, it is the main issue to propound the strengths and weaknesses of each method and arrangement to ensure poorly fit or heavily biased models are not used.

In conclusion, all the models that were trained demonstrate some degree of reliability in accurately predicting risk [12].

DOI: http://dx.doi.org/10.5772/intechopen.1005485

4. Conclusions

The main contribution of this chapter is considering all the characteristics that impact operational performance in a systematic evaluation of safety risk assessment processes while using the machine learning multiclass classification models to predict several levels of classes related to predicting the safety risk assessment over time.

The results of this chapter study also suggest that the application of the machine learning models such as deep learning model and multiclass classification models to safety and quality data not only offers improved data classification but also provides a framework for better safety and quality management systems implementation. Operational companies can employ this model to predict safety risk performance before the operation in order to prevent or decrease any risks of hazards.

Author details

Naimeh Borjalilu
Tehran University, Tehran, Iran

*Address all correspondence to: borjalilu@ut.ac.ir

References

[1] International Civil Aviation Organization. Safety Management Manual (SMM) (Doc 9859 AN/474). 2013. Available from: https://www.icao.int/safety/SafetyManagement/Documents/Doc.9859.3rd%20Edition.alltext.en.pdf

[2] International Civil Aviation Organization. Safety Management, Annex 19 to the Convention on International Civil Aviation. Montreal: ICAO; 2013

[3] van Giffen B, Herhausen D, Fahse T. Overcoming the pitfalls and perils of algorithms: A classification of machine learning biases and mitigation methods. Journal of Business Research. 2022;**144**:93-106. DOI: 10.1016/j.jbusres.2022.01.076

[4] Goerlandt F, Khakzad N, Reniers G. Validity and validation of safety-related quantitative risk analysis: A review. Safety Science. 2017;**99**:127-139

[5] Paltrinieri N, Comfort L, Reniers G. Learning about risk: Machine learning for risk assessment. Safety Science. 2019;**118**:475-486

[6] Faria JM. Machine learning safety: An overview. In: Proceedings of the 26th Safety-Critical Systems Symposium, York, UK. 2018. pp. 6-8

[7] Mohseni S, Wang H, Yu Z, Xiao C, Wang Z, Yadawa J. Practical machine learning safety: A survey and primer. arXiv preprint arXiv:2106.04823, 4. 2021

[8] Ackley JL, Puranik TG, Mavris D. A supervised learning approach for safety event precursor identification in commercial aviation. In: AIAA (American Institute of Aeronautics and Astronautics) Aviation 2020 Forum. 2020. p. 2880

[9] Hegde J, Rokseth B. Applications of machine learning methods for engineering risk assessment—A review. Safety Science. 2020;**122**:104492

[10] Xu Z, Saleh JH. Machine learning for reliability engineering and safety applications: Review of current status and future opportunities. Reliability Engineering & System Safety. 2021;**211**:107530

[11] Torens C, Juenger F, Schirmer S, Schopferer S, Maienschein TD, Dauer JC. Machine learning verification and safety for unmanned aircraft—A literature study. In: AIAA (American Institute of Aeronautics and Astronautics) Scitech 2022 Forum. 2022. p. 1133

[12] Walden H, Hill K, Quinn C, Torres E. Proposal for a Modernized Flight Risk Assessment Tool for General Aviation Pre-Flight Planning. United States: George Mason University; 2022

Chapter 4

Incorporating Risk into Optimization Models

Siddhartha Sampath

Abstract

In modern industry, risk is often understood, communicated, and actioned upon through decision support tools. These tools often incorporate machine learning, statistical and optimization models. These models, especially optimization models, are often point estimate based and must be artfully massaged to incorporate uncertainty to build robust—risk adjusted models that provide a good basis for optimal, long term, goal-based decision making. Probability estimation, simulation and chance constrained optimization are three well known techniques that when used alone or in combination can increase the power of optimization models by considering the underlying risk of the process being optimized when recommending alternatives to decision makers.

Keywords: optimization, risk, chance-constrained, simulation, probability

1. Introduction

People comprehend reality in two primary fashions—an intuitive one and an analytical one [1]. As such these two approaches lend themselves to how we interpret risk in general and how we interpret it in a business context. In an intuitive sense, we tend to associate risk with the fear of experiencing a negative outcome. In a logical sense, we associate risk with negative outcomes and additionally, a probability of that negative outcome occurring. Although the analytical definition of risk is commonly agreed upon, risk is often highly dependent on the specific industry it is referenced in [2]. Furthermore, the perceived "negativity" of an outcome can be subjective, individual decision makers have their own utility functions on how they perceive gains (or losses) [3]. One way to interpret this is through the following thought exercise evaluating the utility of a \$1 gain in two different situations. Suppose you have \$10. An additional dollar is a 10% gain in your total holdings. Now suppose you have \$100. Would \$1 now have the same meaning to you? Probably not. As a motivating factor, \$1 may be less attractive in the second situation than the first. Furthermore, the motivation provided by the potential to gain or lose \$1 may differ between individuals as well.

The second complexity arises from how humans perceive probability—and the aversion to ambiguity as evidenced by the Allais and Elsberg paradoxes [4]. These essentially capture behaviors typically exhibited by decision makers in business where uncertainty is actively avoided despite the potential for larger utility payoffs [5].

Thus, when we talk about "risk" in a business context, the question arises are we talking about the inherent variance associated with the outcomes of a business proposition? Or are we talking about how two different decision makers would arrive at a decision when faced with the same proposition (how they perceive and handle risk)?

It is thus important to properly define risk before proceeding with the rest of this chapter. In order to do this, we borrow the classic definition of a monetary gamble as a set of outcomes $\{x_1, x_2, \ldots, x_n\}$ and a probability distribution $\{p_1, p_2, \ldots, p_n\}$ over $\{x_1, x_2, \ldots, x_n\}$ where p_i denotes the probability of obtaining x_i and $\sum p_i = 1$. The "risk" here is that some of the outcomes in X are undesirable and may occur with a non-zero probability. How decision makers incorporate this information and arrive at a decision, whether using the classic expected-utility framework based on rational choices, prescriptive methods like the analytical hierarchy process or descriptive methods such as prospect theory is beyond the scope of this chapter.

This chapter will instead focus on how to deal with uncertainty in business and three main approaches, probability assessment, simulation, and lastly chance constrained optimization under the assumption that the results of these techniques will be presented to a decision maker via a decision support tool. The decision maker will then use their preferred decision-making strategy to arrive at a decision.

2. Simulation and optimization

Like "risk" the terms "simulation" and "optimization" themselves have many different meanings based on the specific context that they are used in. In general, we will use simulation here to denote a Monte Carlo simulation where many realizations of the random variable as denoted by the monetary gambles we defined earlier are achieved. A thousand sample simulation of a risky gamble X is thus the set of outcomes $\{x_1, x_2, \ldots, x_n\}$ sampled thousand times according to their associated probability of being realized.

Similarly, we define the term optimization in this chapter will denote a constrained optimization problem with constraints, and an objective that must be maximized. A constrained optimization problem can generally be expressed as follows.

$$\text{Max} c(x) \qquad (1)$$
$$\textit{subject to} : a_j(x) \leq b_j \textit{ for } j = 1, \ldots, m$$

One of the simplest examples of a constrained optimization problem is probably the knapsack problem. It can be stated as such.

"A hiker about to embark on a long trek has a knapsack whose weight capacity is W kgs. He has in front of him the option to pick a set of items $1, \ldots, n$. Each of these items has a weight w_i and value v_i associated with it. How can the hiker maximize the value of the items he carries on the trek, without exceeding the weight limit W of the knapsack?"

This can be translated into a binary integer program as follows.

Let $1, \ldots, n$ be N possible items the hiker can choose from.

Let W be the weight limit of the knapsack.

Let w_i be the weight associated with each item and v_i be the value of each item.

We define a *decision variable* x_i which is a binary variable indicating whether the hiker chooses item i or not. In other words, $x_i = 1$ indicates item i was chosen to be included in the knapsack. $x_i = 0$ indicates that the item was excluded.

Solving

$$M1 : Max \sum_i^N v_i x_i \tag{2}$$

$$subject\ to : \sum_i^N w_i x_i \leq W \tag{3}$$

$$x_i \in \{0, 1\} \tag{4}$$

Provides us the optimal set of items that the hiker can choose for a knapsack of weight limit W. We call this the solution vector or solution. The objective term $\sum_i^N v_i x_i$ is maximized and the is the sum of the value of each selected item. The constraint $\sum_i^N w_i x_i \leq W$ ensures that the sum of the weights of the items in the knapsack does not exceed some limit W.

The knapsack problem is a common problem found in many fields including logistics, transportation, warehousing, finance and so on. It can also have many different versions depending on the business context, including having multiple objectives and quadratic constraints. In general, for simpler versions of the problem, many algorithms and solvers exist that can solve large instances ($\sim$ 1 million decision variables) in a reasonable amount of time. The knapsack problem is considered an NP-hard problem, although it scales much better than other NP-hard problems. Many heuristics exist for solving large instances of the knapsack problem to near optimality in polynomial time [6–8].

The model described by M1 is a deterministic model. We use point estimates of the value and the cost to specify the model. What if these estimates were fuzzy instead? What if the weight and value of the items to choose from were random variables that could take on several different values? Based on our previous discussions, we are including a risk into any solution the decision maker specifies. But sometimes, it is not enough to consider the risk of the model coefficients, since we need to evaluate how the model will perform in different scenarios.

To illustrate this, let us modify M1 further with additional complexity. Consider an inventory problem where a manager must decide the quantity of a particular item he must manufacture based on a forecasted demand and a budget. The demand in this sense can be viewed as a random variable. There is a set of possible values that it can assume, with associated probabilities and we assume that any combination of these items will fulfill this overall demand number. If the forecasted demand is higher than the actual demand, the manager will incur an inventory and salvage cost for the unsold items. If the forecasted demand is lower than the actual demand, there is an opportunity cost that the manager will incur. How should the manager approach the problem? This problem can be further complicated by the fact that the value proposition of the item, i.e. the different costs and benefits from manufacturing the item are random variables themselves. The discerning reader will appreciate that even for a single item, this becomes a complicated problem to solve.

Assuming the above problem can be modeled as a knapsack problem, there are two methods adopted in general in industry. One is to simulate the random variable associated with the overall demand, sample different scenarios (percentiles such as 25th, 50th, 90th for example) and then solve the knapsack problem with a different demand. For each solution of the optimization exercise, we can run a simulation again on the output to analyze how each solution would fare for different demand scenarios (keeping in mind that the solution was only optimized for a particular demand scenario). The results of this simulation on the solutions can be compared using appropriate metrics (avg objective value for example) and one of the solutions is chosen.

This is typically referred to as "what-if" scenario analysis [9, 10]. The deterministic version of this problem can be modeled very simply by slightly modifying M1.

$$M2 : Max \sum_{i}^{N} v_i x_i \tag{5}$$

$$subject\ to : \sum_{i}^{N} x_i = F \tag{6}$$

$$\sum_{i}^{N} c_i x_i \leq B \tag{7}$$

$$x_i \in \{0, \ldots, 10\} \tag{8}$$

Here, F is the forecasted demand and the constraint $\sum_i^N x_i = F$ specifies the total quantity of each item adding up to F. However, this F is not the actual demand, which we will denote using D. If F < D, then we assume an opportunity cost of P proportional to the deficit $D - F$. If F > D, we assume that there is an inventory cost of E. But note that we need to solve M2 first and then evaluate it for different values of D. Two additional terms $-\mathrm{P} * \max(\mathrm{D} - \mathrm{F}, 0) - \mathrm{E} * \max(\mathrm{F} - \mathrm{D}, 0)$ can be added to evaluate the true objective. $\sum_i^N c_i x_i \leq B$ is the constraint that ensures that the cost to manufacture all the items does not exceed the budget B. In this problem, each item can be manufactured multiple times and therefore the decision variable x_i can take on any integer value between 1 and 10.

M2 also still assumes point values for v_i and c_i. Solving the model M2 without the $-\mathrm{P} * \max(\mathrm{D} - \mathrm{F}, 0) - \mathrm{E} * \max(\mathrm{F} - \mathrm{D}, 0)$ terms for different values of F (sampled from the distribution for D) will provide different answers. We would then have to evaluate each of these solutions for different values of D sampled from the distribution for D. Ultimately, a decision maker would have to compare the different solutions and choose one. However, none of these solutions would individually be guaranteed to be the "best" solution to satisfy the decision maker's risk appetite. In order to keep it simple, we assume that we would use the mean value of the cost and value associated with each item. **Figure 1** below illustrates this.

For an example of the above problem which is a useful illustration of commonly encountered problems in dealing with risk in industry [11–13], let us define a simple problem instance. Let us assume five products a company is looking to sell. The values associated with these items are normally distributed $N(\mu, \sigma)$ with with the mean and standard deviation as detailed in the table below. Note that the value can assume negative values signifying that the manufacturer assumed a loss on those items.

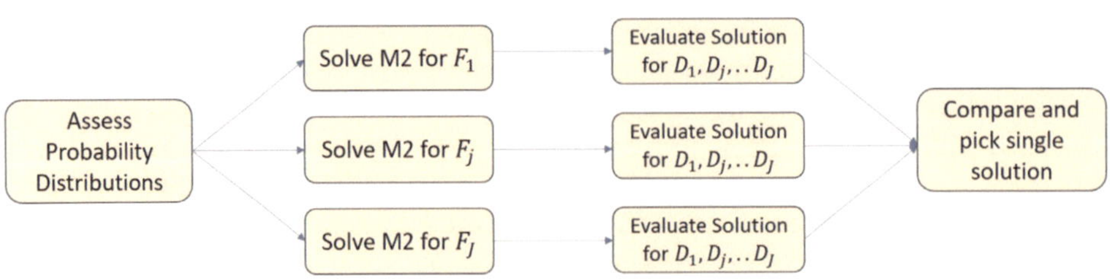

Figure 1.
Solving the knapsack problem with inventory constraints for different forecasted demand values and presenting them to decision makers to choose between them.

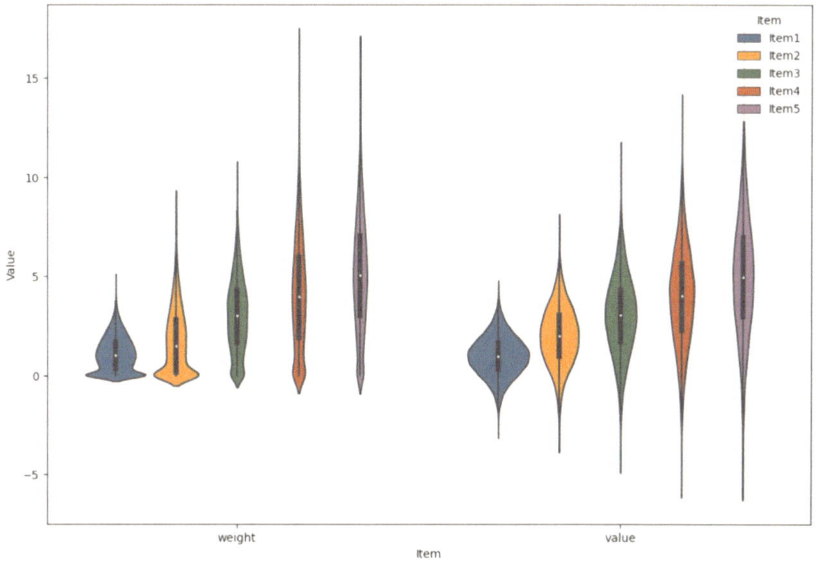

Figure 2.
The probability distributions of the cost(weight) and value of the five items specified in the problem instance depicted using boxplots.

Item	Mean cost	Std cost	Mean value	Std value
A	1	1	1	1
B	1.5	2	2	1.5
C	3	2	3	2
D	4	4	4	2.5
E	5	3	5	3

Table 1.
The mean and standard deviation of the various knapsack items. The costs are modeled using a truncated normal distribution and the benefits are modeled using a standard normal distribution.

We model the costs associated with the manufacture of each item as a truncated normal variable so that we do not incur negative costs as $N_{TRUNC}(\mu, \sigma)$ with the parameters specified in the table below and in **Figure 2** and **Table 1**.

Let us say that, the demand D is a random variable with distribution {5, 10, 15, 20, 25} and the opportunity cost P is distributed {2, 3, 4, 5}, the same as E, with budget of 40.

Assume J samples from D such that $D_1, \ldots, D_j$ and $F_1, \ldots, F_j$ are different values of D and F in each simulation that describes the actual demand and the forecasted demand target to hit. Assume also each simulation has an associated v_{ij}, associated with it, which is the value of the item i in that simulation.

We are now finally at a point where we can play "what-if" scenarios. We could solve the problem for the five values of F, assuming the mean values of $V_i, P\ and\ E$, obtaining five different solutions, then running a secondary simulation on each of these five results for various realized values of D, comparing the five solutions using the average objective value from the simulations, essentially an "evaluation" phase of the generated results. But note the assumptions we are making when solving the problem. Firstly, we are assuming the mean value of v_i in each optimization. We could simulate the values of v_i in the evaluation phase, but we are not inherently considering the riskiness of the item's value in any of the five solutions. Furthermore, we are forced to ignore the $P * \max(D - F, 0)$ and the $E * \max(F - D, 0)$ terms when solving the model since these can only fully be computed during the evaluation phase.

The other method is less common, but one that is gaining traction in industry and that is to run a chance-constrained optimization on the knapsack problem [14–16]. In a chance-constrained optimization model, the simulation is run and is incorporated into the model itself. The chance constrained optimization model then replaces the single constraint in the deterministic case with J different constraints, one for each simulation. We can now pose different conditions on the model to find solutions that satisfy requirements to increase the robustness of the solution to risk in an objective manner, for example solving a chance constrained model can provide a single solution that satisfies conditions like "do not exceed the demand D more than 10% of the time and maximize the mean value of the objective across all simulations" or "do not exceed D more than 50% of the time, but when you do, do not exceed by more than 10% of D and maximize the mean value across all simulations." This in general seems a more robust approach and is easier to communicate to decision makers since it is a single solution that can satisfy a given risk appetite rather than multiple solutions yielded by the first method that a decision maker must choose between. Moreover, it incorporates all aspects of variability of the problem. For example, if there were additional cost attributes associated with the manufacturing of each item that were random variables, this could very simply be added to our chance constrained formulation. **Figure 3** illustrates the general solution procedure describing this process.

For the case where we want 80% probability of the solution does not exceed the budget and maximize the mean value of the solution, we can set up the chance constrained formulation the following way:

$$M3 : Max \sum_{i}^{N}\sum_{j}^{J}\frac{v_{ij}x_i}{J} - \mathrm{P} * \sum_{j}^{J}\frac{y_j}{J} - E * \sum_{j}^{J}\frac{z_j}{J} \tag{9}$$

$$subject\ to : \sum_{i}^{N} x_i = r \tag{10}$$

$$\sum_{i}^{N} c_{ij}x_i - Bb_j \leq B \forall j \in \{1, \dots, J\} \tag{11}$$

$$\sum_{j}^{J} b_j \leq 0.2 * J \tag{12}$$

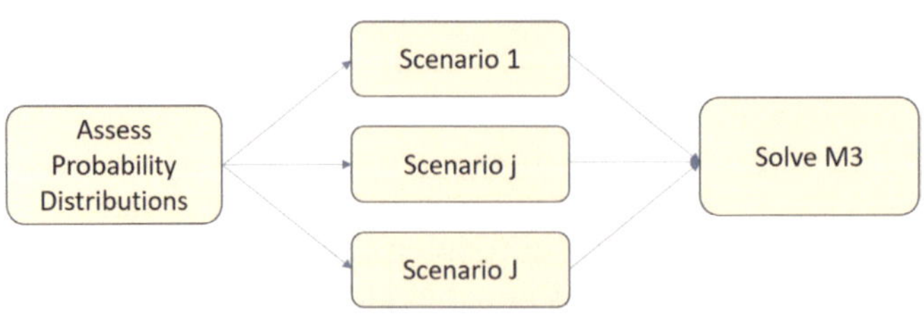

Figure 3.
Solving the knapsack problem with inventory constraints incorporating the risk in value and cost using a chance constrained model. The model produces a single solution.

$$y_j \geq D_j - r - D_{max} u_j \forall j \in \{1, \dots, J\} \tag{13}$$

$$y_j \leq D_{max}(1 - u_j) \forall j \in \{1, \dots, J\} \tag{14}$$

$$z_j \geq r - D_j - D_{max}(1 - u_j) \forall j \in \{1, \dots, J\} \tag{15}$$

$$z_j \leq D_{max} u_j \forall j \in \{1, \dots, J\} \tag{16}$$

$$x_i \in \{0, \dots, 10\} \tag{17}$$

$$r \in \{0, \dots, 50\} \tag{18}$$

$$b_j \in \{0, 1\} \tag{19}$$

$$y_j \in \{0, \dots, \mathrm{D_{max}}\} \tag{20}$$

$$z_j \in \{0, \dots, \mathrm{D_{max}}\} \tag{21}$$

$$u_j \in \{0, 1)\} \tag{22}$$

This model can be explained as follows. To set up M3, we first assume that we have run J simulations where each simulation run specifies a sample value for v_i, c_i and D.

The $\sum_i^N \sum_j^J \frac{v_{ij} x_i}{J}$ term in the objective is the mean value over all simulations. The r variable is how many units we decide to manufacture in total (specified in constraint (10)). Note that we do not need a forecasted demand to manufacture to. Then y_j and z_j are the excess and underage numbers that are equal to $D_j - r$ and 0 respectively if $D_j > r$ and 0 and $r - D_j$ if $r > D_j$. Constraints (13)-(16) specify the values of y_j and z_j ensuring only one of them can be non-zero for any simulation j using binary indicator variable u_j. Constraints (11) and (12) ensures that the budget B can only be exceeded for 20% of the simulations.

Notice the differences between Model M2 and Model M3. In M2, no variability in the costs or value of the items were incorporated. The number of units produced was equal to the forecasted demand. Model M2, would tend to prefer "efficient" items—or items that have a higher value for the mean value of their benefit divided by their cost. M3 however, considers the distribution of cost and benefit values. It also does not assume how many units to produce. This is in fact treated as a decision variable and is an output derived from solving the model.

Solving M2 for a forecasted demand of 10 and a budget of 40 would produce a result that specifies manufacturing three units of item 2 and seven units of item 5. For a forecasted demand of 20, the model recommends 17 units of item 2, 1 unit of item of Item 4 and 2 units of item 5. Which of these is the better solution? The decision maker must evaluate these for different values of the demand and calculate the objective value. If we use the mean values of P and E specified above, the objective value of the two solutions is 16 and 23 respectively. In this case we would pick the second solution where we manufactured 20 units, but what if we were willing to violate our budget 20% of the time? Model M2 does not allow us a framework to easily answer such questions. This is a common case in business where budgets are malleable, and a case can be made for additional funds to manufacture if needed.

In contrast, solving the chance constrained problem gives us a more diversified solution for the problem instance specified of 5 units for item 1, 4 units of item 2, 4 units of item3 and 2 units of item 4 with 20% of the cases violating the budget. The objective value with 100 simulations is approximately 10. Notice how this is significantly lower than either of the two scenarios we modeled using expected values which did not consider the possibility that the combination could exceed the budget and further, ended up overestimating the average objective value. The mean we used in the original problem was replaced by sampling the truncated normal distribution and led to a very different strategy.

Chance constraints are preferred to more complex stochastic formulations of the knapsack problem as they incorporate the results of simulation and do not require a closed form formulation which is difficult with complex distributions like the one used in the problem instance above that involved a truncated normal distribution for the costs of the items [17].

3. Probability assessment

In the discussion above, we assumed that the probability distribution of various random variables describing costs and benefits was known. In the business world, probability assessment plays a crucial role in many different areas by directing resource allocation, risk management techniques, and decision-making procedures. Probability assessment is used in a wide range of industries, including manufacturing, insurance, healthcare, and finance, to measure uncertainty and make decisions. Probabilistic models are used in finance, for example, to predict market trends, evaluate investment risks, and maximize portfolio diversification. Similarly, probabilities are used in medicine to help diagnose conditions, forecast how treatments will work, and assess the efficacy of various medical procedures. Probability assessment is also used in manufacturing and other industries to predict equipment failures, enhance production processes, and reduce downtime. Industry professionals, regardless of the domain, frequently use statistical modeling, expert judgment, and historical data analysis to effectively assess probability. By using an interdisciplinary approach, enterprises may reduce risks, take advantage of opportunities, and predict possible outcomes, which improves overall operational efficiency and competitiveness.

Probability estimation is not a trivial exercise. In many cases, where there is no data available, businesses default to expert opinion to estimate probabilities [18–21]. In fact, the act of estimating accurate probabilities is so important, some companies even hire chief probability officers to measure the inherent riskiness of various activities in the company. As noted by Savage et al., "planning for an uncertain future, calls for a shift in information management–from single numbers to probability distributions–in order to correct the" flaw of averages. "This, in turn, gives rise to the prospect of a Chief Probability Officer to manage the distributions that underlie risk, real portfolios, real options and many other activities in the global economy" [22].

The assumptions behind various probability estimates must be regularly revisited and corrected using statistical analysis based on historical data to ensure optimal decision making.

4. Conclusion

Incorporating risk into optimization models is essential for modern industry decision-making, where uncertainties are prevalent and can significantly impact outcomes. Various techniques for integrating risk considerations into optimization models, including probability estimation, simulation, and chance-constrained optimization can help decision makers make choices that are robust to risk. Probability assessment plays a crucial role in quantifying uncertainties across industries, guiding resource allocation, risk management strategies, and decision-making processes. While empirical data, expert judgment, and statistical analysis are commonly employed for probability estimation, the inherent complexity of uncertainty requires continuous refinement and validation of assumptions to ensure accuracy.

Understanding risk involves both intuitive and analytical perspectives, considering subjective perceptions and objective probabilities of outcomes. Decision-makers interpret risk differently, influenced by individual utility functions and attitudes toward uncertainty. Therefore, defining risk within a business context necessitates careful consideration of industry-specific factors and decision-maker preferences.

Simulation and optimization techniques offer powerful tools for addressing uncertainty in decision-making. Simulation, particularly Monte Carlo simulation, generates multiple realizations of uncertain variables, providing insights into potential outcomes and associated risks. Optimization, on the other hand, identifies optimal decisions under uncertainty, leveraging constraints and objectives to achieve desired outcomes.

Traditional what-if scenarios tend to fall short of providing robust and risk-proof solutions since they present choices that decision makers must still choose between and may not be equipped to do so. This can happen because the solutions analyzed may not truly represent the risk involved since they make assumptions like using point estimates or because the choices must be evaluated using other metrics that decision makers must consider. Often, decision makers may still end up making a subjective decision between the various choices.

Chance constraints provide a more objective method to consider risk by incorporating uncertainty into the model itself. They allow for intuitive questions to be posed and modeled easily and provide a single solution that satisfies various conditions that a decision maker may want from an ideal risk-proof solution. By formulating optimization problems with probabilistic constraints, decision-makers can ensure that solutions meet predefined risk thresholds while maximizing expected outcomes. Chance-constrained optimization offers a unified framework for handling uncertainty, integrating simulation results directly into the optimization process. As illustrated using a knapsack problem, which is a very common problem encountered in industry, considering the distributional characteristics of uncertain parameters, and evaluating solutions across multiple simulations, decision-makers can make informed choices that balance risk and reward effectively. By leveraging probability estimation, simulation, and chance-constrained optimization techniques, industries can mitigate risks, exploit opportunities, and enhance operational efficiency, ultimately achieving their strategic objectives in a competitive landscape.

Acknowledgements

Many thanks to my former director at Intel, Karl G Kempf for the many discussions around this topic that led to me shaping my views. Thanks must be given also to Esma Gel, John Fowler and Jorge Sefair for their advice on the topic during my stint at ASU.

Author details

Siddhartha Sampath
Meta, Newark, CA, USA

*Address all correspondence to: siddsam@gmail.com

References

[1] Epstein S. Integration of the cognitive and the psychodynamic unconscious. American Psychologist. 1994;**49**:709-724

[2] Pablo AL. Managerial risk interpretations: Does industry make a difference? Journal of Managerial Psychology. 1999;**14**(2):92-108. DOI: 10.1108/02683949910255142

[3] Crundwell FK. Decision tree analysis and utility theory. Finance for engineers: Evaluation and funding of capital projects. 2008:427-456

[4] Munthe C. The price of precaution and the ethics of risk. Springer Science & Business Media; 19 May 2011

[5] Zappia C. Leonard savage, the Ellsberg paradox, and the debate on subjective probabilities: Evidence from the archives. Journal of the History of Economic Thought. 2021;**43**(2):169-192

[6] Pferschy U, Schauer J, Thielen C. Approximating the product knapsack problem. Optimization Letters. 2021; **15**(8):2529-2540

[7] Baldo A et al. The polynomial robust knapsack problem. European Journal of Operational Research. 2023;**305**(3): 1424-1434

[8] Faenza Y, Segev D, Zhang L. Approximation algorithms for the generalized incremental knapsack problem. Mathematical Programming. 2023;**198**(1):27-83

[9] Arsham H et al. Sensitivity analysis and the "what if" problem in simulation analysis. Mathematical and Computer Modelling. 1989;**12**(2):193-219

[10] Bisogno S et al. Combining modelling and simulation approaches: How to measure performance of business processes. Business Process Management Journal. 2016;**22**(1):56-74

[11] Cappanera P, Trubian M. A local-search-based heuristic for the demand-constrained multidimensional knapsack problem. INFORMS Journal on Computing. 2005; **17**(1):82-98

[12] Garcia-Guarin J et al. Schedule optimization in a smart microgrid considering demand response constraints. Energies. 2020;**13**(17):4567

[13] Jian C, Wu Y, Li M. Energy efficient allocation of virtual machines in cloud computing environments based on demand forecast. In: Advances in Grid and Pervasive Computing: 7th International Conference, GPC 2012, Hong Kong, China, May 11–13, 2012. Proceedings 7. Berlin Heidelberg: Springer. p. 2012

[14] Song Y, Luedtke JR, Küçükyavuz S. Chance-constrained binary packing problems. INFORMS Journal on Computing. 2014;**26**(4):735-747

[15] Li X, Liu S, Wang J, Chen X, Ong YS, Tang K. Data-Driven Chance-Constrained Multiple-Choice Knapsack Problem: Model, Algorithms, and Applications. 26 Jun 2023. arXiv preprint arXiv:2306.14690

[16] Küçükyavuz S, Jiang R. Chance-constrained optimization: A review of mixed-integer conic formulations and applications. 21 Jan 2021;**2**. arXiv preprint arXiv:2101.08746

[17] Li P, Arellano-Garcia H, Wozny G. Chance constrained programming approach to process optimization under

uncertainty. Computers & Chemical Engineering. 2008;**32**(1–2):25-45

[18] Ouchi F. A Literature Review on the Use of Expert Opinion in Probabilistic Risk Analysis. Policy Research Working Paper 3201. World Bank. 2004. Available from: http://econ.worldbank.org/

[19] Ramler R, Felderer M. Experiences from an initial study on risk probability estimation based on expert opinion. In: 2013 Joint Conference of the 23rd International Workshop on Software Measurement and the 8th International Conference on Software Process and Product Measurement. IEEE. 23 Oct 2013. pp. 93-97

[20] Clemen RT, Winkler RL. Combining probability distributions from experts in risk analysis. Risk Analysis. 1999;**19**: 187-203

[21] Sampath S, Gel ES, Kempf KG, Fowler JW. A generalized decision support framework for large-scale project portfolio decisions. Decision Sciences. 2022;**53**(6):1024-1047

[22] Savage S, Scholtes S, Zweidler D. Probability management: Planning for an uncertain future calls for a shift in information management–from single numbers to probability distributions–in order to correct the "flaw of averages." This, in turn, gives rise to the prospect of a chief probability officer to manage the distributions that underlie risk, real portfolios, real options and many other activities in the global economy. OR/MS Today. 2006;**33**(1):20-28

Chapter 5

Radon Release Features of Different Uranium Mines and the Relative Public Effective Doses

Lechang Xu, Hui Zhang, Yalan Wang, Jie Niu and Xiangnan Dai

Abstract

This chapter attempts to introduce the development of a radon sampling device for the investigation of the return air shaft of mines. Also, different radon measurement methods were compared, which were based on the investigation of the radon release and diffusion pattern in the return air shaft, alkaline (CO_2 + O_2) *in situ* leaching uranium and sulfuric acid *in situ* leaching uranium, as well as analysis on continuous distribution characteristics of environmental radon activity concentration. Additionally, comparative studies on the normalized radon release of various mining processes and aerosol radionuclide distribution of various uranium mining facilities were implemented. Finally, the radiation dose prediction evaluation and cumulative radiation dose evaluation based on effluent monitoring and environmental monitoring, respectively, together with normalized public dose evaluation are also presented.

Keywords: radon, uranium mines, return air shaft, *in situ* leaching, public radiation dose, normalized radiation dose, distribution characteristics

1. Introduction

Radon is known as a natural carcinogen. The release of radon during uranium mining has always been considered as a potential radioactive health hazard, and it is one of the most concerned radionuclides in uranium mining, which needs to be monitored and controlled.

El-Fawal established a joint solution model of radon, radon daughter, air volume, and wind pressure in mine ventilation network, and realized the prediction of radon activity concentration and radon daughter concentration in mine tunnel and working surface [1]. Sahu et al. reviewed various sources of radon in underground uranium mines (such as orebodies, filling tailings, crushed ore, and mine water) and factors affecting the radon release process on the basis of investigation on the effects of internal factors such as ore grade, ^{226}Ra content, water content, material porosity, and surface area, and external factors such as atmospheric pressure, temperature and ventilation on radon activity concentration in uranium mines [2]. Long Huijia et al. carried out studies on the distribution of radon activity concentration and wind speed at the outlet of a uranium mine ventilation shaft by adopting the single-point sampling method combined with the interpolation and integration program [3]. Mudd implemented research on the radon release from uranium mining and milling in Australia in depth, and the cumulative radon release was analyzed and compared with the UNSCEAR method [4].

Most studies on the migration and diffusion of radionuclides in the atmosphere focus on the impact of leakage of nuclear power plants or nuclear installations on the environment and human beings and the evaluation of the consequences of nuclear accidents. The study on the impact of radionuclides in the exhaust gas of uranium mine on human beings and the environment is mainly based on field measurement, combining with numerical simulation. Xie Dong et al. conducted a numerical simulation study on the relationship between radon activity concentration and diffusion distance in the exhaust gas of uranium mine [5]. Wang Wenxian et al. conducted MATLAB simulation of atmospheric stability and wind speed on radon diffusion from exhaust releases of underground uranium mines [6]. Ye Yongjun et al. numerically simulated the distribution of surface radon activity concentration within 1 km around the return air shaft of uranium mines by the Gaussian plume model and analyzed the effects of initial radon activity concentration and air volume from the wellhead of return air shaft [7].

Li Xiaojun investigated and monitored the radon activity concentration and radon release rate into the ambient air of the *in situ* leaching mine, and grasped the characteristics of radon release with an *in situ* leaching uranium mine in Xinjiang. UAIR-FINE software was used to simulate the distribution of radon activity concentration, individual dose, and collective dose in each sub-region within 20 km around the leaching pool and evaporation pond [8]. In terms of the current problem of lack of measurement methods for radon release from recovery wells of *in situ* leaching field, Zhang Hui et al. calculated the characteristics of radon activity concentration release by using a theoretical model and verified it with field measured data, and found the normalized radon release of *in situ* leaching was only about a quarter of that of underground mining [9].

In China, uranium mining and milling industry was founded in 1958. Since the establishment of China's uranium mining and milling industry, China has established underground mines, open-pit mines, and *in situ* leaching mines, and developed natural uranium extraction processes such as stirred leaching, pile leaching, *in situ* blasting leaching, and *in situ* leaching, among which the *in situ* leaching process includes sulfuric acid leaching and $CO_2 + O_2$ alkaline leaching, which is the only country in the world that covers various types of uranium mining and milling processes. China's uranium mining and milling facilities have gone through three stages of development: operation, shutdown, and long-term stewardship of post decommissioning. In the process of production and operation of these uranium mining and milling facilities, to different extent the surrounding environment has been affected. In 1986, the first radiation environmental quality assessment of China's nuclear industry over the past 30 years was organized. Due to the absence of shutdown and decommissioning of uranium mines, a comprehensive investigation and evaluation of the current radiation environment of uranium mines and milling plants in operation was only carried out. The evaluation results show that the collective dose resulting from uranium mining and milling to the public accounts for 91.5% of the entire nuclear fuel cycle system, and gaseous pathway dose in uranium mining and milling facilities accounts for 76% of total dose of gas and liquid pathway, in which ^{222}Rn and short-lived daughters account for 92% of the gaseous pathway [10]. It can be inferred that ^{222}Rn is the main cause for radiation environmental impact.

Since 1990s, uranium mines have experienced shutdown, decommissioning and newly built in China. In order to compare the radiation environmental effects of uranium mining and milling facilities with different uranium mining/milling processes and different stages including in operation and shut-down stages, a comprehensive and systematic investigation and evaluation were carried out.

2. Research content and methods

Underground mine return air shaft is the largest radon release source. The concentration is usually tens to hundreds of thousands Becquerel per cubic meter. The return air shaft wind speed is usually more than 10 m/s, humidity up to 98%, and temperature over 40°C. Therefore, the conventional radon monitoring device is not workable in the underground mine environment, and its release characteristics and its migration and diffusion characteristics in this environment have not yet been studied. For *in situ* leaching, there is no liquid pathway of public exposure by adoption of evaporation pool, and radon migration is the main public exposure pathway. Radon mainly comes from *in situ* leaching solution. A large amount of radon will be released during the changes of pressure, temperature, and disturbance. However, for the systematic studies of radon release and diffusion, key facility source terms have not been carried out. Due to the rapid change of environmental radon activity concentration with spatial and temporal variation, the research on the migration and diffusion of radon in return air shafts and *in situ* leaching uranium mining facilities need to carry out large-scale simultaneous distribution monitoring. Therefore, the monitoring technology of radon migration and diffusion of return air shaft is the primary problem. The challenges are to study and reveal the characteristics of release, migration, and diffusion of gas-borne pollutants.

In addition, the distribution of main radionuclides in aerosols is not clear according to previous studies. The relative distribution of the radionuclides ^{238}U, ^{226}Ra, ^{210}Pb, and ^{210}Po in aerosols is of common concern and needs to be revealed.

2.1 Radon monitoring method

On the basis of the available measurement technology for radon, the short-time, long-time, and continuous radon measurement technologies were combined by the authors. With help of a multi-directional simultaneous monitoring, the spatial and temporal distribution of radon and its daughter nuclides could be studied around uranium mines. Monitoring technology of radon release and migration of non *in situ* leaching uranium mines is represented by the maximum source term of return air shaft of underground uranium mine, and refers to the distribution characteristics of non-point source boundary of waste rock heap and tailings pond. The radon diffusion and migration characteristics in an *in situ* leaching uranium mine is revealed by the measurement of the radon in the liquid transport pipe, the uranium-rich lixiviate pool, the milling plant, the evaporation pool, and the surrounding environment.

The main pollutants in uranium mines are radon and aerosols, mainly ^{238}U, ^{226}Ra, ^{222}Rn, ^{210}Pb, ^{210}Po, etc., including return air shaft, uranium-rich lixiviate pool, and nonpoint sources such as tailings pond and waste rock debris. Radon activity concentration monitoring is affected by a variety of factors, including measurement methods and devices, measurement point and time period, facilities and surrounding environmental conditions, etc.

Cumulative radon measurement refers to the way in which the radon activity concentration is characterized by cumulative mean radioactivity response over a period of time. The cumulative measurements include long-time and short-time radon measurements, represented by solid track etching and electret methods, respectively. The short-time measurement method refers to a cumulative detection time of more than 4 hours but no more than a half-life of radon. It monitors the fluctuations of the radon activity concentration. On the other side, long-time monitoring shows the integral activity over the whole measuring period. Short-time and long-time cumulative radon measuring techniques are complemented by a continuous instantaneous radon monitoring. The radon monitoring in this study applied these radon monitoring methods.

Passive radon measurement is mainly adopted in the selection of method to reduce the environmental disturbance caused by the sampling process, in which environmental disturbance could lead to the wrong judgment of radon activity concentration levels at different points. The electret radon measurement method was used as the short-time cumulative radon measurement method, and the solid-track radon detection method was used as the long-time cumulative radon measurement method.

Table 1 shows the radon monitoring methods of different facilities according to investigation and comparison with the characteristics of various types of facilities.

2.2 Monitoring technology of radon release, migration, and diffusion in the return air shaft

2.2.1 Monitoring technology and device of radon release in the return air shaft

The monitoring technology of the characteristics of radon release and surrounding migration and diffusion distribution of underground uranium mining mainly applies to the return air shaft facilities. The return air shaft radon measuring device was developed considering the flow field and flow velocity characteristics of the return air shaft facilities. The cumulative radon activity concentration distribution was monitored simultaneously at different flow field locations in the return air shaft. In the return air shaft around a 1000-m range, we used the short-time and long-time cumulative radon monitoring to investigate the radon migration and diffusion distribution characteristics at the same time.

At the stable site of the 3.6-m diameter return air shaft, the parameters of wind speed, wind pressure, humidity, and stability are measured at the radial 0.11, 0.34, 0.62, 1.02, 1.75, 2.49, 2.88, 3.16, 3.40 m to determine the characteristics of radon discharge field of the return air shaft. Computational formula:

$$I = 0.16 \times (Re) - 0.125 \quad (1)$$

$$Re = v d \rho / \eta \quad (2)$$

where is: I: turbulence strength; Re: Reynolds number; v: average velocity; d: hydraulic diameter; ρ: medium density; η: medium dynamic viscosity coefficient. Turbulent intensities of less than 1% were classified as "low turbulent intensity." Turbulent intensities higher than 10% were considered as "high turbulent intensities."

Radon activity concentration detectors were placed at different depths and at different points in the radial direction to monitor the radon activity concentration in the return air shaft and at the exhaust outlet. At the same time, the radon activity

Monitoring object	Monitoring method	Monitoring device	Monitoring characteristics
Return air shaft	Electrets method	The E-perm electret	Short-time cumulative measurement
	Solid-track etching method	KF606B radon and γ personal dosimeters	Long-time cumulative measurements
Lixiviate preparation pool vent of *in situ* leaching	Solid-track etching method	KF606B radon and γ personal dosimeters	Long-time cumulative measurements
	Electrets method	The E-perm electret	Short-time cumulative measurement
	Electrostatic collection method	RAD7,PQ2000, SARAD1688	Instantaneous radon measurements
Milling plant (except lixiviate preparation pool) vent of *in situ* leaching	Solid-track etching method	KF606B radon and γ personal dosimeters	Long-time cumulative measurements
Uranium-rich lixiviate	The flicker chamber method	FD 125, Radon thorium analyzer	Instantaneous radon measurements
	Electrostatic collection method	RAD7,PQ2000, SARAD1688	Instantaneous radon measurements
Environmental radon	Electrets method	The E-perm electret	Short-time cumulative measurement
	Solid-track etching method	KF606B radon and γ personal dosimeters	Long-time cumulative measurements
	Electrostatic collection method	RAD7,PQ2000, SARAD1688	Instantaneous radon measurement, continuous radon measurement
Water surface radon flux	Solid-track etching method	KF606B radon and γ personal dosimeters	Long-time cumulative measurements
Ground radon flux	Electrostatic collection method	RAD7,PQ2000, SARAD1688	Instantaneous radon measurement, continuous radon measurement

Table 1.
Methods and device for radon monitoring.

concentration of the flow field in the return shaft, near the exit and the outlet, is monitored. Radon monitoring device were handled with a bracket structure and lifting system (see **Figures 1** and 2).

A bracket system supports the monitoring devices and equipment. A lifting system is used to ensure the effective operation of the bracket system. The designed working positions are 0, −0.5, and − 1 m, measured from the shaft head. Two mutually vertical trusses fix the detectors on the upper two layers. Two monitoring trusses are set on the cross −1 m layer, and an aerosol sampler bracket is placed in the center to collect aerosol samples. The monitoring truss is arranged at the shafthead depth of 0, −0.5, and −1 m. Nine monitoring points are arranged on the truss with distance from the center of 9, 20, 40, 60, 80, and 100%. About 19 simultaneous monitoring data points are arranged in the same batch. The monitoring devices includes solid track detectors and electret detectors. Field operation is shown in **Figure 3**.

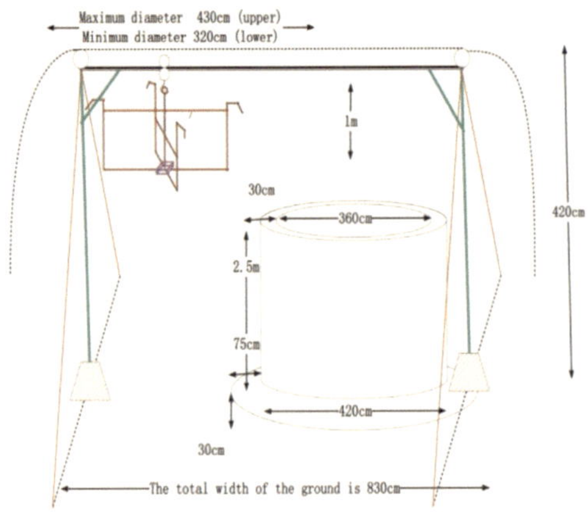

Figure 1.
Radon measuring device sketch in return air shaft.

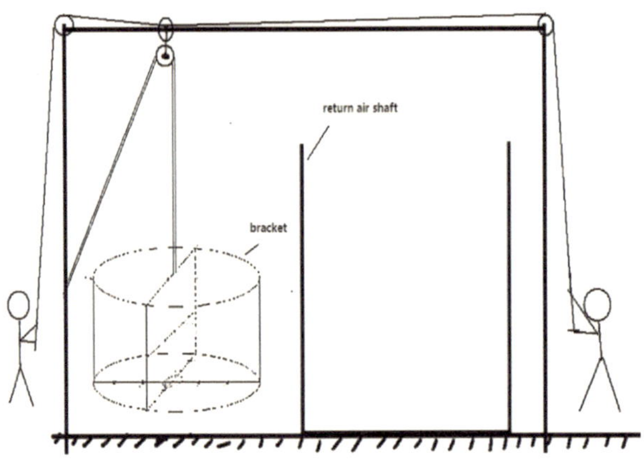

Figure 2.
Schematic diagram of device operation.

Figure 3.
Arrangement of cumulative radon detector.

The distribution of the radon activity concentration and flow field in the shaft is found out by measuring the radon activity concentration and wind speed in the shafthead at many points. A multi-point-interpolation-integral estimation method for investigation of the radon release in the return air shaft is established. The radon release concentration is calculated by integrating the change of the flow field and concentration of each monitoring point. Each radon activity concentration monitoring point controls the radon activity concentration flux of certain fan or in the field of fan column range, and calculates the radon activity concentration flow in the return air shaft. The following formulas are given:

$$P = \sum_{i=1}^{n} v_i C_i S_i \tag{3}$$

$$v_i = \frac{1}{x_{i+1} - x_i} \int_i^{i+1} v d_x \tag{4}$$

$$C_i = \frac{1}{x_{i+1} - x_i} \int_i^{i+1} C d_x \tag{5}$$

$$S_i = \frac{\pi \left(r_i^2 - r_{i-1}^2 \right)}{2} \tag{6}$$

where V_i is the mean wind speed in the control range of the concentration monitoring point i (m/s). C_i is the radon activity concentration at the control point i (Bq/m^3). S_i is the interface control area for the control point i (m^2). P is the release rate (Bq/s). According to the release rate and the release time t (s), it can be calculated as annual release Q (Bq/a).

$$Q = P \cdot t \tag{7}$$

Comprehensive formula (3) ~ formula (7), available as follows:

$$Q = \sum_{i=1}^{n} \frac{1}{x_{i+1} - x_i} \int_i^{i+1} v d_x \cdot \frac{1}{x_{i+1} - x_i} \int_i^{i+1} C d_x \cdot \frac{\pi \left(r_i^2 - r_{i-1}^2 \right)}{2} \cdot t \tag{8}$$

2.2.2 Monitoring technology of the distribution characteristics of radon migration and diffusion around the return air shaft

The study on the distribution of radon around the return air shaft mainly adopts the combination of long-time and short-time cumulative radon monitoring, supplemented by the continuous monitoring of instantaneous radon-monitoring. The short-time radon measuring points are placed in 16 directions around the return air shaft. The main measuring points of long-time cumulative radon are placed at the dominant wind direction ENE and downwind direction WSW, the minimum wind frequency direction SE and downwind direction NW, and the six directions, in distances of 30, 50, 100, 200, 300, 500, 800, and 1000 m. About 10% parallel sample points were set to verify the quality of the monitoring data. Short-time radon monitoring period was between 8 and 48 h: The long-time radon monitoring was set to 90 days per batch (1 quarter/time, totaling 4 times). Instantaneous radon continuous monitoring was set for 72 hours. Both the operation and the shutdown phase were monitored. Temperature, the relative humidity, air pressure, wind direction, and wind speed were recorded during the monitoring process.

2.3 Monitoring technology of radon release, migration, and diffusion of *in situ* leaching uranium

The monitoring of the release, migration, and diffusion of radon at the *in situ* leaching uranium site includes the monitoring of the uranium-rich lixiviate pool, the uranium milling plant, the liquid transport pipe, the drilling, the evaporation pool, etc.

2.3.1 Sampling and monitoring of radon in liquid transport pipes

In the process of *in situ* leaching, the uranium-rich lixiviate of the ore-bearing layer is extracted to the surface from dozens of meters to several meters underground. More radon is dissolved under the underground pressure. As the groundwater is pumped to the surface, the pressure decreases, and a large amount of radon is released. It is also lost when the water flows during the passage through the transport pipe. In addition to its migration with water and airflow, radon can also be transferred in two phases.

Radon gas in liquid transport pipes is released into the air after contact with the atmosphere after entering the uranium-rich lixiviate pool. Radon gas that is not released in the pool will continue to flow into the plant and return to the underground. This process is a dynamic process, and such high concentrations cannot easily be monitored. Therefore, two invention patents, namely, the free gas radon-collecting device and the free gas radon-monitoring device, were adopted (see **Figures 4** and **5**). Radon-containing gas in the extraction bottle was introduced into the scintillation chamber and radon activity concentration was measured and calculated.

2.3.2 Calculation of radon activity concentrations in the recovery well

In this study, a one-dimensional vertical radon diffusion model was used to theoretically simulate the radon diffusion concentration. The process of radon migration and diffusion in the recovery well depends on the radon-containing extraction solution in the gas-phase interface, and the radon migration and diffusion into the upper air, connecting with the air through the gap of the top hole of the recovery well. The vertical constant temperature diffusion experiment of radon is designed to determine the effective diffusion coefficient of radon release in the recovery well environment and estimate the radon activity concentration at the recovery well outlet at different depths and different water radon activity concentrations. An experimental bench device for one-dimensional vertical diffusion of radon is built.

Figure 6 shows a schematic diagram of the device structure. It includes radon source, diffusion pipe, thermostatic casing, radon measurement device, and end

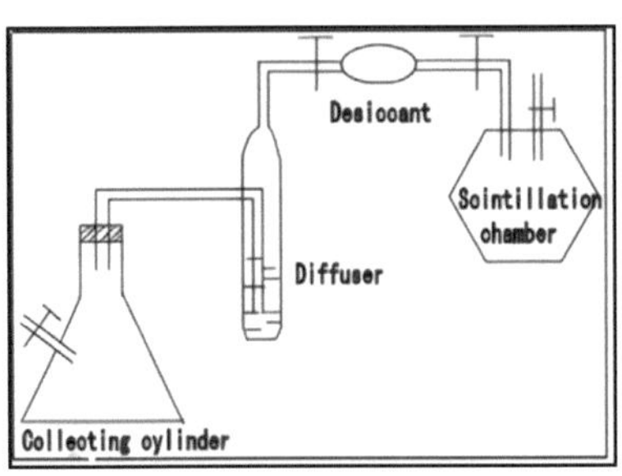

Figure 4.
Schematic diagram of infusion tube gas gathering device.

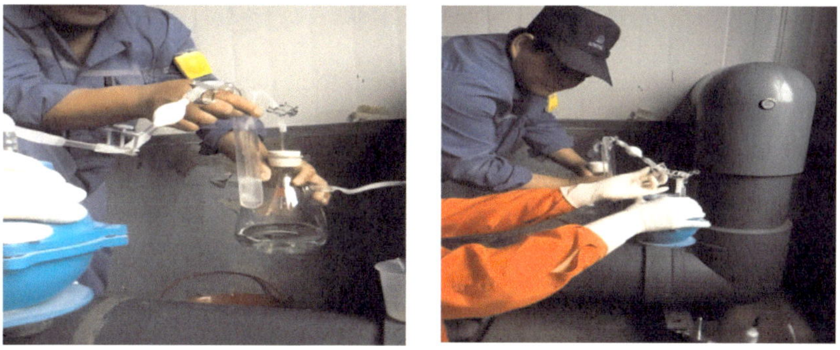

Figure 5.
Gas and radon sampling site of infusion tub device.

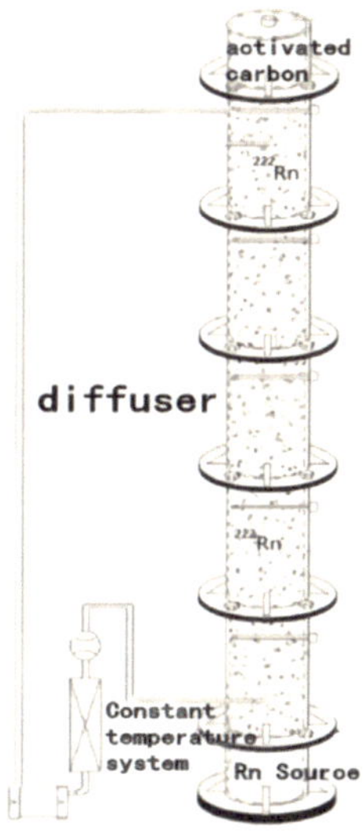

Figure 6.
Schematic representation of the one-dimensional diffusion experimental setup.

radon absorption device. The diffusion behavior of radon gas in the extraction pipe belongs to the diffusion migration distribution of infinite vertical space, which behaves conforming to Fick's theorem and is similar to the distribution of release gas in uniform non-radioactive media of layered ore body. The diffusion distribution can be described by the following equation:

$$N = N_0 e^{-\sqrt{\lambda/D_e}\cdot x} \tag{9}$$

where N is the radon gas concentration from the space to the radon source x, x is the distance of a certain point in the pipe from the radon source, and λ is the decay constant of ^{222}Rn, D_e is the effective diffusion coefficient of radon gas in the experimental environment, and N_0 is the radon gas concentration close to the location of the radon source point.

2.3.3 Monitoring and calculation of radon flux in the evaporation pool

A device for measuring water surface Radon flux is developed as shown in **Figure 7**. Passive diffusion electret is used to measure radon. The electret at the bottom of the air-collecting hood is installed and places the air-collecting hood on the surface of the measured medium to ensure good edge air tightness. At 4 to 10 h of accumulation, an electret is replaced and the measurement is completed after 20 h. Another electret was used to measure the radon activity concentration in the surface air. About 3 ~ 5 radon activity concentration data in different time periods were obtained by the effective decay constant method.

The calculation method of measured Radon flux includes slope method and effective decay constant method. Where the slope method is an approximation method, which assumes $\lambda t << 1$, C_0 is 0 (λ is the radon decay constant, t is the measurement time, and C_0 is the initial radon activity concentration) approximately replaces the Radon flux curve. The inherent relative error between the measured value and the true value of the slope method is $(1-e^{-\lambda t})/(\lambda t) - 1$. For this purpose, when the accumulation time exceeds 10 h its inherent system error is at least 5% using this method. Because the device adopts the radon measurement method and the short-time electret, the accumulation time should be at least 4 h, so the device adopts the effective decay constant method for the calculation of the radon flux.

The calculation method is performed as follows:

$$E_e = \frac{V\lambda_e\left(C - C_0 e^{-\lambda_e T}\right)}{S\left(1 - e^{-\lambda_e T}\right)} \tag{10}$$

where Ee is the effective Radon flux; V is the volume of cumulative radon cover; S is the area of the cumulative radon cover; λe is the effective decay constant, which is related to the air leakage coefficient designed by the device, and the properties of the

Figure 7.
Schematic diagram of measuring device (left) and structure (right).

measured uniform medium; T is the measurement period; C is the measured radon activity concentration; C_0 is the initial radon activity concentration, usually is 0.

According to formula (10), the Radon flux can be calculated as long as λe is known. Let xi be the n radon activity concentration data with equal time period T and yi be the i radon activity concentration data with equal time period T, where xi is not equal to yi. Then:

$$\lambda e = \frac{1}{T} \ln Y \frac{1}{b} Y \tag{11}$$

$$b = \frac{\sum x_i y_i - \frac{1}{n}\left(\sum x_i \sum y_i\right)}{\sum x_i^2 - \frac{1}{n}\left(\sum x_i\right)^2} = e^{-\lambda_e T} \tag{12}$$

$$a = \frac{\sum y_i}{n} - b\frac{\sum x_i}{n} = \frac{E_e S}{\lambda_e V}\left(1 - e^{-\lambda_e T}\right) \tag{13}$$

$$E_e = \frac{V\lambda_e a}{S(1-b)} \tag{14}$$

2.3.4 Monitoring technology of the distribution characteristics of radon migration and diffusion around the uranium-rich lixiviate pool and milling plant

The radon activity concentration around the uranium-rich lixiviate pool of *in situ* leaching mine was investigated by short-time cumulative radon-detectors. In the maximum wind frequency (up and down direction), minimum wind frequency (up and down direction), middle wind frequency (up and down direction), 11 distances of 2, 5, 10, 30, 30, 50, 100, 200, 300, 500, 800, and 1000 m were chosen for a 24-h monitoring.

At the larger and lower wind frequencies (up and down direction) of 50, 100, 300, and 500 m around the plant, namely four directions and four distances, the solid nuclear track detectors were exposed for the same time period.

3. Study results

3.1 Rules of radon release, migration, and diffusion in the return air shaft

3.1.1 Distribution law of radon activity concentration in the return air shaft

Solid nuclear track cumulative radon detectors were used to monitor the distribution of radon activity concentration in the return air shaft. The analysis of the data showed that the radon activity concentration at the center of radius was lower than at the edge and lower than at the center at both depths of −1 and −0.5 m in the return air shaft. However, the radon activity concentration at the outlet of the return air shaft showed high radon activity concentrations at the shaft's center of radius (see **Figures 8–10**).

The radon activity concentration in the return air shaft, measured by KF-606B solid track cumulative radon detectors, showed a radon activity concentration range of 21.0–63.7 kBq/m^3 and a mean value of 37.5 kBq/m^3. The radon monitored by electrets showed a radon activity concentration range and a mean value of 18.4–127 kBq/m^3 and 47.6 kBq/m^3, respectively. The two methods were comparable.

It is recommended to use the sampling points used in this study to achieve the optimal sampling effect and representativeness. Under appropriate conditions, at least

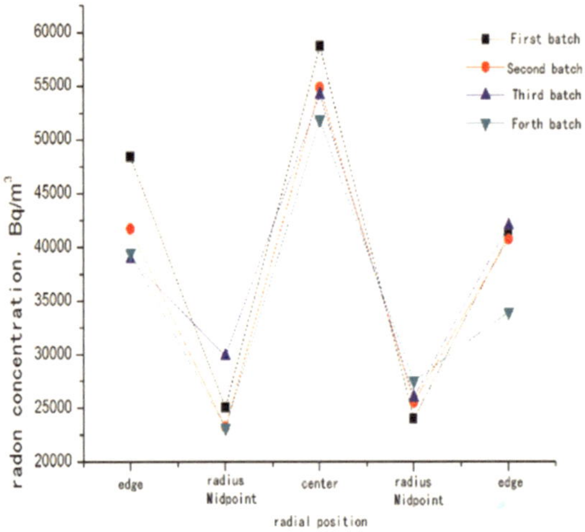

Figure 8.
Distribution of radon activity concentration in X and Y at a depth of −1 m.

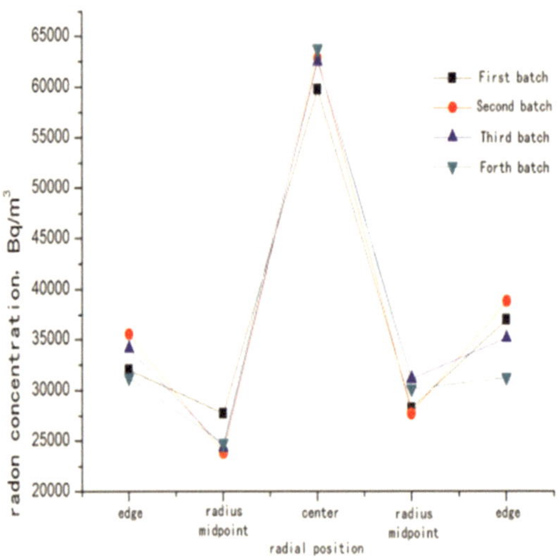

Figure 9.
Distribution of radon activity concentration at a depth of −0.5 m.

one sampling point can be set at the center point of −1 m to obtain trusted radon activity concentration level data. It can be simplified as the return air shaft outlet of −1 m and the monitoring of radon activity concentration release level at the radial 1/3. The monitoring method could be a passive radon monitoring, while solid track detectors are recommended.

3.1.2 Distribution law of radon migration and diffusion in the environment around the return air shaft

From June 2014 to May 2015, a long-time radon monitoring study on the distribution law of radon activity concentration around the return air shaft was conducted in four batches for different directions and distances: two batches under normal

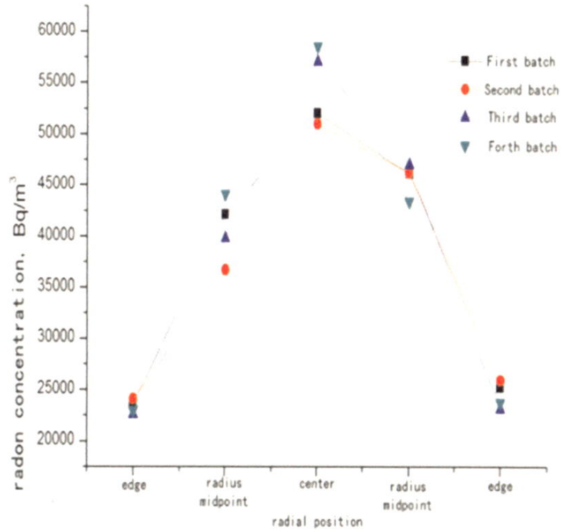

Figure 10.
Distribution of radon activity concentration at the return air shaft exit (0 m).

production conditions from June to November 2014 and two batches at shutdown mode of the mine, from December 2014 to May 2015. Three blank samples were retained in each batch for background monitoring. The three blank background was 31, 27, and 26 Bq/m^3.

For the short-time cumulative radon, the data were processed by the synergistic Kriging method (Co-Kriging), and a contour map of the radon activity concentration distribution was drawn. According to the meteorological data in 2014 of the nearest weather station within 50 km of the return air shaft, the distribution of radon activity concentration within 1000 m around the shaft was estimated by mode estimation. **Figures 11** and **12** show the contour plots of the radon activity concentration within 1000 m around the shaft under operation condition and shutdown condition, respectively. At **Figure 11**, the mine was under the operation conditions, and the mean radon activity concentration of the return air shaft was 37 kBq/m^3. The radon activity concentration outside a distance of 400 m from the shaft was about 75 Bq/m^3. Furthermore, the radon contribution outside a distance of 500 m did not exceed 30 Bq/m^3. The radon contribution outside 800 m did not exceed 20 Bq/m^3, which is within the local background level. The diffusion of radon around the 15# return air shaft of J mine was mainly affected by the dominant wind direction, and a "sink" was formed at about 100 m southwest of the return air shaft, with the maximum concentration of 1732 Bq/m^3. The radon activity concentration reduced to less than 100 Bq/m^3 at about 400 m around the return air shaft, and the radon activity concentration outside the 500 m range was basically lower than 50 Bq/m^3.

Figure 12 shows the radon distribution at operation and shutdown mode of the mine. The average radon activity concentration in the N, ENE, SE, and NW directions decreased by 70–97 Bq/m^3, and the average radon activity concentration in the S, WSW directions was lower than at operation mode (330–373 Bq/m^3). Compared with the shutdown status, the radon activity concentration outside the 500 m perimeter changed little. The boundary range of their influence on the radon in the surrounding environment is about 500 m to 800 m under operation and shutdown mode.

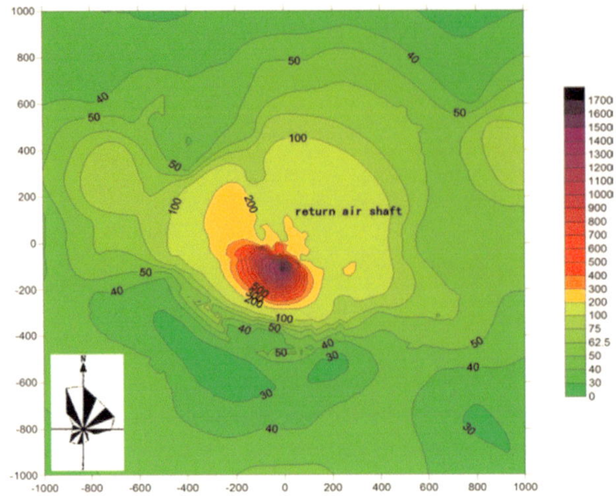

Figure 11.
Mine in operation mode. Radon-contour map.

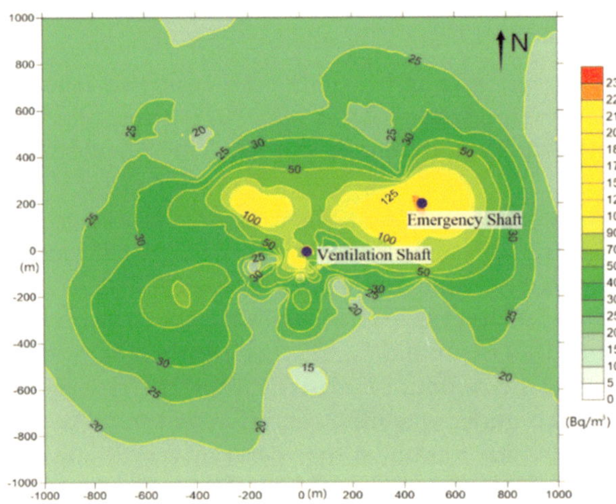

Figure 12.
Mine at shutdown mode. Radon-contour map.

The distribution of the radon activity concentration in the WSW and ENE directions of the return air shaft had obvious distance distribution characteristics. **Figures 13** and **14** show the overall behavior of the radon activity concentration decreased with distance. The maximum radon activity concentration in this direction occurs 100 m away from the return air shaft, while the radon activity concentration in the return air shaft at 500 m decreased to the local background level.

Under the shutdown status of the mine, the radon activity concentration decreased, and the average at the outlet was 850 Bq/m^3, generally lower than 100 Bq/m^3 outside the air shaft. Among them, the radon activity concentrations outside 500 m were generally lower than 30 Bq/m^3. A high concentration area appeared 400 m east-northeast of the return air shaft, as there was a uranium mine safety

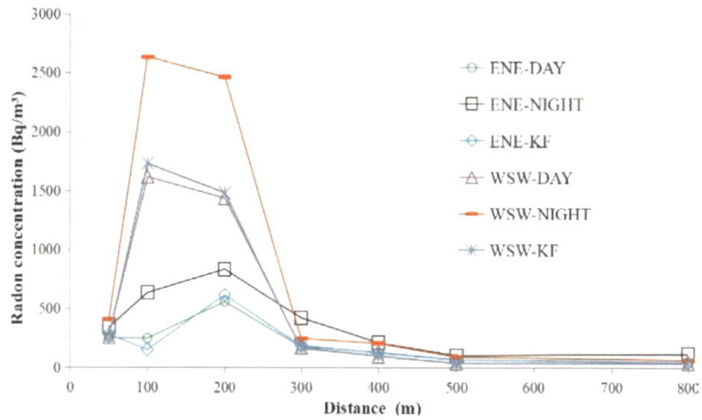

Figure 13.
Radon activity concentrations in the ENE and WSW directions at different distances and monitoring times.

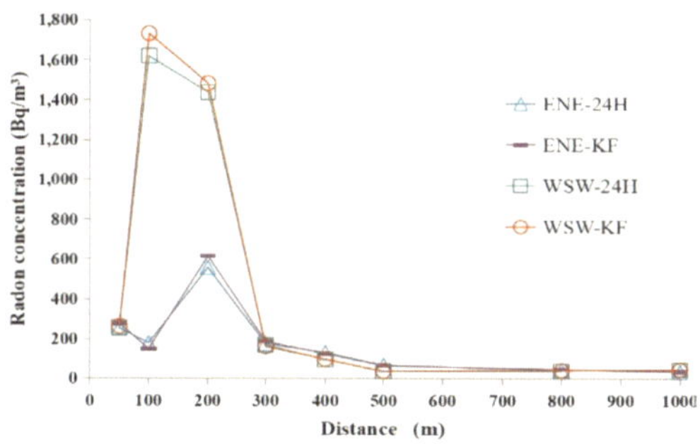

Figure 14.
Comparison of the long-term and short-term monitoring data in the ENE and WSW directions.

measure shaft at the bottom of the normal production state, and the unorganized air flow from the roadway in the shutdown state due to the change of the wind pressure at the bottom, resulting in an increase of radon. This phenomenon proved the effectiveness of the distribution scheme of radon activity concentration distribution in this study.

3.2 Rules of radon release, migration, and diffusion of the *in situ* leaching uranium mine

3.2.1 Investigation of the radon release from the uranium-rich lixiviate pool

Short-time cumulative radon monitoring was used for the survey of the radon activity concentration at the uranium-rich lixiviate pool (passive sampling integral measurement).

The monitoring locations are shown in **Figure 15**. The results are shown in **Table 2**. The average concentration of the radon release in the uranium-rich lixiviate pool was 65.9 kBq/m^3. The radon activity concentration decreased rapidly at 2 m distance from the opening of the uranium-rich lixiviate pool, with an average concentration of about 2400 Bq/m^3. It is recommended to seal the uranium-rich lixiviate pool.

Figure 15.
Electret monitoring position at the opening of uranium-rich lixiviate pool.

For the convenience of calculation, assuming that at the opening of the open uranium-rich lixiviate pool, the amount of radon release in the uranium-rich lixiviate pool can be estimated by the following equation:

$$Q = 0.25CRnvAT\sum_{x=1}^{x} P(x)ABS(\sin x) \tag{15}$$

Q is the annual radon release (Bq); C_{Rn} is the calm wind mean radon activity concentration (Bq/m^3, set 47.7 kBq/m^3); V is the average annual long-time wind speed (m/s, set 3.6 m/s in this survey); A is the opening area (m^2, set 8m^2); T is the time of 1 year; ABS is absolute; P is the wind frequency; x is the wind direction angle. Then, the annual release of radon in the uranium-rich lixiviate pool was 6.47 TBq/a.

The first phase uranium-rich lixiviate pool of T Mine was sealed in November 2016 and in September 2018. The radon activity concentration at the sealed uranium-rich lixiviate pool and its surrounding areas was measured with RAD 7-detectors, which were distributed along the downwind direction of the main wind direction. Results are shown in **Table 3**. As can be seen, the radon activity concentration close to the pool wall of the sealed uranium-rich lixiviate pool was 4.2 kBq/m^3, which was more than one order of magnitude lower of radon activity concentration of the unsealed uranium-rich lixiviate pool. The sealing of the uranium-rich lixiviate pool had a significant effect on the radon release. The radon release of the trap was reduced by 95%, and the whole process release of the *in situ* leaching was reduced by 1–2 orders of magnitude.

3.2.2 Investigation of radon release in the exhaust outlet of alkaline (CO_2 + O_2) in situ *leaching uranium milling plant*

The alkaline (CO_2 + O_2) *in situ* leaching refers to lixiviate formed by adding gaseous carbon dioxide and oxygen to native groundwater. The milling plant includes the main plant and a supporting plant. The main plant is separated from the supporting plant by walls. The main workshop includes an ion exchange area and a precipitation area, and the supporting workshop includes a storage pool area (qualified liquid pool, shower pool, lean liquid pool), an uranium-rich lixiviate filtration area, a reverse osmosis area, a recoil area, a liquid distribution pool, etc. The storage area and the distribution pool are sealed spaces, with only an exhaust outlet. There are 13 exhaust vents for the active ventilation facilities in the milling plant (see **Figure 16**). The cumulative radon activity concentration of each exhaust outlet was monitored and the annual radon release was calculated as shown in **Table 2**.

Parameter	Uranium-rich lixiviate pool	Milling plant							Wellfield drilling	Evaporation pool
		Lixiviate preparation pool	Leaching filter unit	Qualified lixiviate pool	Reverse osmosis area	Settling zone	Rejection zone	Chemical analysis unit		
Mean radon activity concentration, Bq/m^3	65,582	52,690	1423	1012	1438	2117	1711	497	257	45
Radon release amount, GBq/a	6074	3859	70.34	16.47	23.40	32.92	27.84	0.86	30.83	19.01
Percentage	59.8	37.97	0.69	0.19	0.23	0.33	0.27	0.01	0.30	0.20

Table 2.
Radon activity concentration and annual radon release in uranium-rich lixiviate pool, milling plant, wellfield drilling and evaporation pool of $CO_2 + O_2$ in situ *leaching T mine.*

State of the uranium-rich lixiviate pool	Radon activity concentration at the wall of the sump and downwind, Bq/m^3						
	Pool wall	2 m	5 m	100 m	500 m	800 m	1000 m
Before sealing	46,000	6179	2712	315	85	49	39
After sealing	4230	376	257	101	52.7	26.7	13.5

Table 3.
Comparison of radon activity concentration around uranium-rich lixiviate pool of in situ *leaching.*

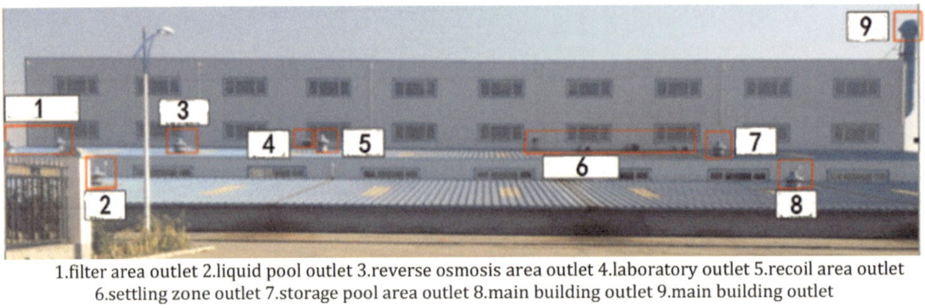

Figure 16.
Distribution of each exhaust outlet of milling plant of T uranium mine. 1. Filter area outlet. 2. Liquid pool outlet. 3. Reverse osmosis area outlet. 4. Laboratory outlet. 5. Recoil area outlet. 6. Settling zone outlet. 7. Storage pool area outlet. 8. Main building outlet. 9. Main building outlet.

3.2.3 Investigation of radon release in evaporation pool and recovery well and liquid transport pipe

3.2.3.1 Investigation of radon release in the evaporation pool

The electret radon flux-measuring device was used to float at the edge of the 0.5 m of the evaporation pool, and the release rate of radon was measured 25 m away from the inlet pipe of the evaporation pool. After freezing in winter, the radon flux of 5 points was measured in the center of the evaporation pool and in the diagonal position of the evaporation pool. The radon release rate is shown in **Figure 17**. The monitoring

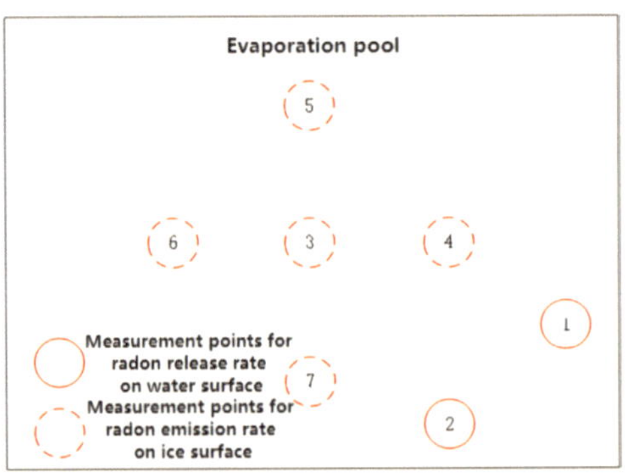

Figure 17.
Schematic diagram of radon release rate on ice surface.

result of radon flux was that the radon flux was monitored for each water surface point and 5 ice surface points, and the monitoring data of radon flux of 7 water surface and 25 ice surface points were obtained. The mean value of water surface radon flux was 0.06 Bq/m^2·s, and the mean radon flux on the ice surface was 0.035 Bq/m^2·s. According to the area of the evaporation pool of 12160m^2 calculated, the icing period of no. 1 evaporation pool is 5 months, 12,160 m^2 × 8760 h × 3600 s (0.060 Bq/m^2·s × 7/12 + 0.035 Bq/m^2·s × 5/12) = 19.01 GBq.

3.2.3.2 Investigation of radon release from recovery wells

The effective diffusion coefficient was calculated according to the fitting of the experimental results of the recovery well, and the radon activity concentration at the outlet of the recovery well could be calculated according to the formula. The fit is shown in **Figure 18**. The results using the least squares method can calculate the radon effective diffusion coefficient De. It was 0.0578 cm^2·s^{-1} at 16°C in vertical space. The radon activity concentration at the outlet of the recovery well was compared with the monitoring of the radon activity concentration. It is known that the radon activity concentration in the mining area is 2 × 10^5 Bq/m^3. At 16°C, the equilibrium Oswald coefficient of radon in the gas-liquid two phases is 0.3, and the air radon activity concentration at the *in situ* leaching liquid level is 60 kBq/m^3. The distance between the borehole level of the stope is 8–10 m, and the mean value is 9.06 m. The radon activity concentration at the outlet of the extraction borehole was calculated to be 254.4 Bq/m^3.

3.2.3.3 Investigation of radon release in the liquid transport pipe

According to the measurement of water radon and gas radon in the seven centralized control rooms of T uranium mine, the mean activity concentration of gas radon in the total valve of the central control chamber was 75.5 and 2076 Bq/L, respectively, totaling 2152 Bq/L. The radon activity concentration of leaching solution lifted from the ground was reduced by about 70%, that was, by 1472 Bq/L in surface liquid distribution pool after the recovery process, resulting in the water radon activity concentration 680 Bq/L.

Gaseous radon and dissolved radon in liquid transport pipes are typical radon sources *in situ* leaching uranium mines. Gaseous radon contributed significantly to the increased activity concentration and release of radon in the non-closed liquid trap.

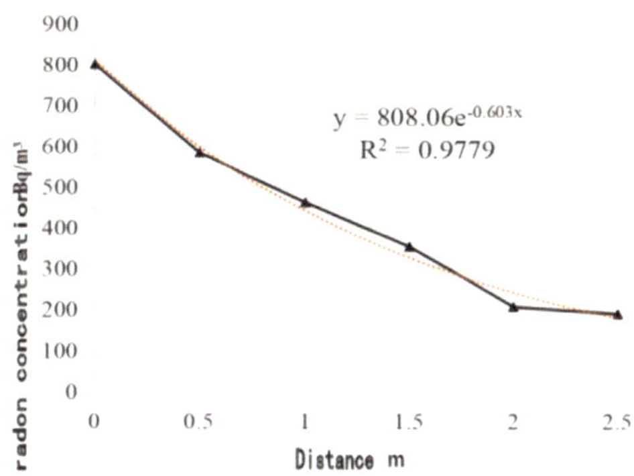

Figure 18.
Fitting results of radon activity concentration and diffusion distance.

3.2.4 Investigation on the migration and diffusion law of radon in the surrounding environment uranium-rich lixiviate pool of in situ *leaching facilities*

The migration and diffusion of radon around the uranium-rich lixiviate pool is mainly affected by the dominant wind direction, as shown in **Table 4**. As the wind direction was SW on the day, the maximum was 6179 Bq/m^3 at a distance of 2 m in the NNE direction. The radon activity concentration around the uranium-rich lixiviate pool in T mine was basically lower than 50 Bq/m^3 outside the 800 m range. The radon activity concentration of the X1 mine by acid method was lower than 40 Bq/m^3 outside the 500 m range and the outdoor background value of 9.76–41.48 Bq/m^3 basically consistent, indicating that the environmental impact of radon release in the open uranium-rich lixiviate pool generally did not exceed 500 m.

3.2.5 Investigation on the migration and diffusion law of radon in the surrounding environment of the in situ *leaching uranium milling plant*

As can be seen in **Table 4**, radon activity concentration around milling plant was lower than around the uranium-rich lixiviate pool. The radon activity concentration decreased with the increasing distance from the milling plant. The environmental impact of radon release generally did not exceed 500 m.

3.3 Analysis of normalized radon release in various types of uranium mines

At present, only J uranium mine is conventional mixing leaching in underground mining mines in China, while the others are heap leaching. The release of radon is small. The normalized radon release amount in each mine is shown in **Table 5**. The release of normalized radon from underground mining is 0.327 $\sim$ 4.372 TBq/tU, while the normalized release of radon from *in situ* leaching uranium is only 0.0852 TBq/tU. L mine has the largest normalized radon release with 4.372 TBq/tU, followed by DB uranium mine with 3.912 TBq/tU, both deposits using *in situ* blasting leaching process. In 1978, the Pacific Northwest Research Institute measured and estimated the normalized radon release of 18 return air shafts, and its value was (0.205 $\sim$ 12.69) TBq/tU. The normalized radon release of return air shafts in China was lower than that of return air shafts in the United States. After removing the *in situ* blasting leaching, the normalized radon release of underground mines in China is 0.63 $\sim$ 2.83TBq/tU, which is generally higher than that reported in UNSCEAR 2000 by 0.018 $\sim$ 2.36 TBq/tU (1 $\sim$ 2000G/tU_3O_8) [11]. Mainly, China's uranium resource endowment is poor, and the underground mining scale is small and scattered. The radon of underground milling mainly comes from the return air shaft, which accounts for 97.29% of the radon release amount of the whole underground mine. The amount of normalized radon released on *in situ* leaching uranium is significantly lower than that of underground uranium, which is about 4.27% of the amount of underground radon released.

3.4 Distribution pattern of aerosol radionuclides

The aerosol radioactivity levels in the source terms and nearby atmospheric environment of various uranium mining and milling facilities are shown in **Table 6**.

As can be seen in **Table 6**, ^{238}U concentration in the aerosol from the operational underground uranium mining is higher, significantly higher than the shutdown and *in situ* leaching facilities. The high point of ^{238}U concentration is the shafthead, which

***In situ* leaching**	**Radon activity concentration distribution around the uranium-rich lixiviate pool**											**Distribution of radon activity concentration around milling plant**				
	Distance/m	**2**	**5**	**10**	**30**	**50**	**100**	**300**	**500**	**800**	**1000**	**Factory boundary**	**50**	**100**	**300**	**500**
$CO_2 + O_2$ *in situ* leaching uranium T mine	NNE	6179	2712	1232	1675	772	315	174	85	49	39	75	59	51	54	39
	SSW	264	251	87	85	85	83	60	43	42	40	646	548	145	79	56
	WNW	561	203	153	112	94	110	87	60	43	43	300	180	54	23	37
	ESE	3115	1898	1124	426	238	209	91	87	40	43	71	62	52	68	41
	N	143	97	102	101	99	87	53	42	42	36	—	—	—	—	—
	S	158	143	118	103	83	91	59	43	43	32	—	—	—	—	—
	W	2154	4318	285	131	94	84	43	43	43	31	—	—	—	—	—
	E	4187	2183	1246	2281	1238	310	184	86	60	49	—	—	—	—	—
Acid *in situ* leaching uranium X1 mine	E	381	—	—	—	35	34	41	24	—	—	691	575	122	81	27
	W	258	—	—	—	32	68	45	26	—	—	327	189	54	22	28
	N	151	—	—	—	46	49	38	31	—	—	—	—	—	—	—
	S	115	—	—	—	33	40	42	28	—	—	—	—	—	—	—
	SE	—	—	—	—	16	45	41	40	—	—	75	65	52	71.4	23
	NW	—	—	—	—	48	24	25	35	—	—	87	61	53	56	31

Table 4.
Radon activity concentration distribution in uranium-rich lixiviate pools, milling plant (Bq/m^3).

Installation	Conventional underground mining								*In situ* blasting and leaching of underground mining				*In situ* leaching mining		
	B mine	Q mine	C mine	J mine	M mine	DC mine	Weighted mean	Mean ratio, %	L mine	DB mine	Weighted mean	Mean ratio, %	$CO_2 + O_2$ *in situ* leaching T mine	Acid *in situ* leaching X1 mine	Acid *in situ* leaching X2 mine
Return air shaft	0.63	0.22	1.15	0.63	1.42	2.83	1.08	96.26	4.34	3.81	4.15	98.67			
Tailings pond	0.041	0.066	0.016	0.014	0.044	0.015	0.029	2.58	0.013	0.068	0.032	0.76			
Heap leaching site	0.006	0.041	0.018	0.005	0.018	0.008	0.013	1.16	0.019	0.034	0.024	0.57			
In situ leaching site													0.085	0.029	0.040
Total	0.67	0.327	1.184	0.649	1.482	2.853	1.122	100	**4.372**	**3.912**	4.206	100	0.085	0.029	0.040

Table 5.
Normalized radon release (TBq/tU) in each mine (shaft).

Facilities type		^{238}U		^{226}Ra		^{210}Pb		^{210}Po	
		Scope	Mean	Scope	Mean	Scope	Mean	Scope	mean
Source term	Underground mine in operation	0.0198 ~ 33.6	29.3	0.0099 ~ 0.143	0.041	0.78 ~ 294	53.6	0.14 ~ 37.0	2.62
	In situ leaching	0.038 ~ 0.7	0.174	0.0056 ~ 0.05	0.021	0.29 ~1.87	0.61	0.042 ~ 3.45	0.92
	Mine shutdown	0.0077 ~ 6.5	0.66	0.0039 ~ 0.034	0.013	0.33 ~3.57	1.36	0.15 ~ 1.83	0.66
Environment	Underground mine in operation	0.011 ~ 1.4	0.748	0.0062 ~ 0.1	0.021	0.59 ~ 10.43	4.32	0.089 ~ 2.82	1.04
	In situ leaching mine	0.014 ~ 0.041	0.027	0.0063 ~ 0.03	0.016	0.83 ~ 4.38	2.03	0.142 ~ 1.37	0.49
	Mine shutdown	0.0046 ~ 6.5	0.488	0.003 ~ 0.053	0.017	0.09 ~ 5.44	1.66	0.13 ~ 1.90	0.96
	Domestic	0.0059 ~ 0.044	0.023	0.0047 ~ 0.12	0.021	0.11 ~ 9.4	1.70	0.03 ~ 1.84	0.82

Table 6.
Aerosol radioactivity levels in different mine sources and environments (mBq/m^3).

leads to higher ^{238}U concentration resulting from the diffusion of uranium dust. The concentration of radionuclide in the *in situ* leaching mine is low. In most of the aerosol samples in the surrounding environment of various uranium mines, the radioactivity of ^{210}Pb and ^{210}Po was slightly higher than the UNSCEAR 2000 [11] recommended reference value (0.5 and 0.05 mBq/m^3).

In aerosols, the activity concentrations of ^{210}Pb and ^{210}Po are mostly higher than that of ^{238}U and ^{226}Ra and show the following row: ^{210}Pb > ^{210}Po > ^{238}U > ^{226}Ra. In a few cases, the ^{238}U activity is higher than the ^{210}Po-activity. The radionuclide (except radon) activity concentrations of aerosol from gaseous source term and environment are 6–10 and 4–6 orders of magnitude lower than the radon activity concentration, respectively. In the gas phase, the activity concentrations of ^{210}Pb and ^{210}Po in aerosols from other release sources (such as tailings ponds, waste rock yards, hydrometallurgy plants) except the shaft head are comparable with nearby residential sites. The average activity concentrations of ^{210}Pb and ^{210}Po in the shaft aerosol are 1 to 2 orders of magnitude higher than that of nearby settlements. ^{210}Pb and ^{210}Po in the uranium shafthead aerosol should be paid attention to.

3.5 Predictive evaluation of effective doses for the public based on the effluent

3.5.1 Expansion of the atmospheric evaluation model

AERMOD is selected as the application model to simulate and predict atmospheric migration and diffusion of radionuclides from uranium mines. A dose evaluation is carried out in combination with China's Y30AIR radiation dose evaluation software, thus overcoming the shortcomings of AERMOD's [12] non-dose evaluation function. Y30AIR's non-point source integration, terrain and complex meteorological processing function, and synthesizing are the advantages of both models.

3.5.2 Prediction and evaluation results

From the radiation environmental impact assessment results of different types of facilities, the yearly emissions of different mine types are shown in **Table 7**. Public doses of critical resident groups (all contributed by gaseous pathway) from various mines are presented in **Table 8**. **Table 9** shows the individual effective doses distribution of critical groups of various facilities. The collective doses from the gas phase and the liquid phase are shown in **Table 10**. **Table 11** presents the comparison of the impact of radiation environment of various facilities. Finally, the collective effective doses of the public are given in **Table 12**.

Conclusions from **Tables 8–12** are follows:

Mine type	Pathway of irradiation	Release amount (Bq/a)							
		Rn-222	U-238	U-234	Ra-226	Th-230	Po-210	Pb-210	total
Conventional mining in operation	Gas	4.47×10^{14}	8.94×10^{10}	4.60×10^{8}	4.38×10^{8}	4.60×10^{8}	4.38×10^{8}	4.38×10^{8}	4.47×10^{14}
	Liquid	—	3.59×10^{9}	3.59×10^{9}	1.33×10^{9}	7.10×10^{8}	1.19×10^{9}	1.92×10^{9}	1.23×10^{10}
	Total	4.47×10^{14}	9.30×10^{10}	4.05×10^{9}	1.76×10^{9}	1.17×10^{9}	1.63×10^{9}	2.36×10^{9}	4.47×10^{14}
In situ leaching mining	Gas	3.36×10^{13}	—	—	—	—	—	—	3.36×10^{13}
	Liquid	—	—	—	—	—	—	—	—
	Total	3.36×10^{13}	—	—	—	—	—	—	3.36×10^{13}
Shutdown mining	Gas	2.61×10^{14}	—	—	—	—	—	—	2.61×10^{14}
	Liquid	—	3.30×10^{9}	3.04×10^{9}	1.56×10^{8}	4.37×10^{8}	1.47×10^{7}	2.26×10^{7}	6.97×10^{9}
	Total	2.61×10^{14}	3.30×10^{9}	3.04×10^{9}	1.56×10^{8}	4.37×10^{8}	1.47×10^{7}	2.26×10^{7}	2.61×10^{14}
Decommissioning mining operation	Gas	4.59×10^{11}	—	—	—	—	—	—	4.59×10^{11}
	Liquid	—	3.92×10^{6}	3.92×10^{6}	3.13×10^{6}	2.31×10^{6}	9.03×10^{5}	2.18×10^{6}	—
	Total	4.59×10^{11}	3.92×10^{6}	3.92×10^{6}	3.13×10^{6}	2.31×10^{6}	9.03×10^{5}	2.18×10^{6}	4.59×10^{11}
Total	gas	7.42×10^{14}	8.94×10^{10}	4.60×10^{8}	4.38×10^{8}	4.60×10^{8}	4.38×10^{8}	4.38×10^{8}	7.42×10^{14}
	Liquid	—	6.89×10^{9}	6.63×10^{9}	1.49×10^{9}	1.15×10^{9}	1.21×10^{9}	1.95×10^{9}	1.93×10^{10}
	Total	7.42×10^{14}	9.63×10^{10}	7.09×10^{9}	1.92×10^{9}	1.61×10^{9}	1.65×10^{9}	2.39×10^{9}	7.42×10^{14}

Table 7.
Releases of gaseous and liquid radioactive effluent.

Critical resident groups and the main contributing sources		Conventional mining in in operation				*In situ* leaching mining			Shutdown mining							Decommissioned
		M Mine	J Mine	L Mine	Q Mine	X1 Mine	X2 mine	T mine	B Mine	H Plant	DC Mine	JH Mine	JR Mine		TC Mine	LC Mine
													JR-1	JR-2		
Critical residents group	Position	S	NE	SE	W	NW	ENE	SW	WNW	ENE	W	ESE	WNW	N	SES	W
	Distance (km)	0.5	5.39	0.85	0.51	1.44	5.46	1.50	2.00	0.9	0.63	1.88	1.6	1.2	1.12	0.37
	Dose value (mSv/a)	0.33	0.48	0.94	0.002	0.016	0.017	0.004	0.006	0.39	0.014	0.024	0.027	0.028	0.076	0.008
Main contribution source terms		Cotton pit south return air shaft	Shannan waste rock debris	102 Return air shaft	Return air shaft	Wellfield, milling plant	Tower A uranium-rich lixiviate pool	Lixiviate preparation pool	Tailings pond	Tailings pond	Tailings pond	708 tailings pond	Tailings pond	Tailings pond e	Well field	Tailings pond

Table 8.
Critical groups of uranium mines and their effective dose based on effluents (key pathways are gaseous and key radionuclide is ^{222}Rn).

Type	The ratio of public maximum individual dose distribution (%)				
	<0.1	≥0.1, < 0.25	≥0.25, < 0.5	≥0.5, < 1	≥1
Conventional mining in operation	25	0	50	25	0
In situ leaching mining	100	0	0	0	0
Shutdown mining	75	12.5	12.5	0	0
Decommissioning mining	100	0	0	0	0
Total	64.3	7.1	21.4	7.1	0

Table 9.
Individual dose distribution in critical resident groups of various facilities.

1. From the perspective of the total amount of source release, the percentage of releases of conventional operational mines, *in situ* leaching mines, closure and decommissioned mines are 60.2, 4.5, 35.2, and 0.1% of the total released amounts in uranium mining and milling facilities, respectively. The contribution of conventional mines in operation mode is the largest (**Table 7**).

2. Among all types of mines, the maximum individual effective dose for the public caused by gaseous effluent is mainly the conventional in-operation mine. The *in situ* blasting leaching mine (L Mine) gives the largest dose (0.94 mSv/a). The decommissioning LC Mine gives the smallest dose (0.008 mSv/a). The key radionuclide of the gaseous effluent of all mines is ^{222}Rn. The maximum individual effective dose for the public due to liquid effluent is mainly in conventional in-operation mines, with a dose value of 1.84×10^{-2} mSv/a. Here, the key radionuclide is ^{210}Pb. The critical residential groups caused by each mine are the adjacent mine residential points, with the key radionuclide ^{222}Rn. The main pathway is by inhalation. During normal operation of some mine, the radiation effect on the surrounding living public is close to or even beyond the dose constraint value (0.5 mSv/a). Due to the large amount of source term release from nearby waste rock debris with numerous facilities in the region, the dose of critical resident group is close to the constraint value in J Mine (**Table 8**).

3. From the maximum value of public individual dose based on effluent evaluation, the main contribution is less than 0.1 mSv/a; the public dose caused by *in situ* leaching and decommissioned mines is less than 0.1 mSv/a, showing a small radiation impact on the public (**Table 9**).

4. Routine in-operation mines are the facility type with the greatest impact on the surrounding environment and the public. L Mine have relatively large influence, mainly related to the high grade of the deposit, the large source intensity, and the proximity of the settlement and facilities; among the closed facilities, the main contribution facility is 272 plant tailings pond, which is related to its large scale. It has not been decommissioned yet. Although the source term of *in situ* leaching mine is relatively large, its dose contribution to the public is relatively small due to its low emission and sparse surrounding population; the radiation impact of the decommissioning mines is minimal, and its source term and the public dose are lower than those of other types of mines.

Type	Pathway of irradiation	Collective Dose (person Sv/a)							
		Rn-222	U-238	U-234	Ra-226	Th-230	Po-210	Pb-210	total
Conventional mining in in operation	Gas	3.67×10^{1}	4.09×10^{-2}	4.79×10^{-2}	4.79×10^{-2}	2.17×10^{-1}	2.93×10^{-3}	1.49×10^{-2}	3.70×10^{1}
	Liquid	—	2.49×10^{-3}	7.19×10^{-4}	3.83×10^{-3}	1.74×10^{-4}	4.46×10^{-3}	6.31×10^{-2}	7.48×10^{-2}
	Total	3.67×10^{1}	4.34×10^{-2}	4.86×10^{-2}	5.17×10^{-2}	2.17×10^{-1}	7.39×10^{-3}	7.80×10^{-2}	3.71×10^{1}
In situ leaching mining	Gas	6.45×10^{-1}	—	—	—	—	—	—	6.45×10^{-1}
	Liquid	—	—	—	—	—	—	—	—
	Total	6.45×10^{-1}	—	—	—	—	—	—	6.45×10^{-1}
Shutdown mining	Gas	1.94×10^{1}	—	—	—	—	—	—	1.94×10^{1}
	Liquid	—	3.87×10^{-4}	4.15×10^{-4}	1.17×10^{-2}	4.03×10^{-4}	2.48×10^{-3}	3.94×10^{-2}	5.48×10^{-2}
	Total	1.94×10^{1}	3.87×10^{-4}	4.15×10^{-4}	1.17×10^{-2}	4.03×10^{-4}	2.48×10^{-3}	3.94×10^{-2}	1.94×10^{1}
Decommissioning mining	Gas	2.00×10^{-2}	—	—	—	—	—	—	2.00×10^{-2}
	Liquid	—	2.04×10^{-7}	1.95×10^{-7}	3.49×10^{-8}	4.19×10^{-8}	4.02×10^{-14}	7.46×10^{-11}	4.76×10^{-7}
	Total	2.00×10^{-2}	2.04×10^{-7}	1.95×10^{-7}	3.49×10^{-8}	4.19×10^{-8}	4.02×10^{-14}	7.46×10^{-11}	2.00×10^{-2}
Total	Gas	5.67×10^{1}	4.09×10^{-2}	4.79×10^{-2}	4.79×10^{-2}	2.17×10^{-1}	2.93×10^{-3}	1.49×10^{-2}	5.71×10^{1}
	Liquid	—	2.88×10^{-3}	1.13×10^{-3}	1.56×10^{-2}	5.77×10^{-4}	6.94×10^{-3}	1.03×10^{-1}	1.30×10^{-1}
	Total	5.67×10^{1}	4.38×10^{-2}	4.90×10^{-2}	6.35×10^{-2}	2.18×10^{-1}	9.87×10^{-3}	1.17×10^{-1}	5.72×10^{1}

Table 10.
Collective dose distribution of air-liquid effluent.

5. The total public collective dose caused by the whole uranium mining and milling system is 57.2 person·Sv/a, ^{222}Rn is the main contributor to the collective dose (98.9%); main pathway is by air (99.8%). Underground mines in operation and closure modes largely contribute to the collective dose (49.5 and 49.6%, respectively). This is related to the large release of the two kinds of mines, with their contributions of 60.2 and 35.2%, respectively.

Mine type	Total source term (Bq/a)	Maximum contribution of the public individual dose (mSv/a)		Public collective dose (person·Sv/a)
		Based on the effluents	Based on the environmental monitoring data	
Conventional mining in operation	4.47×10^{14}	0.94	0.72	3.71×10^{1}
In situ leaching mining	3.36×10^{13}	0.017	0.28	6.45×10^{-1}
Shutdown mining	2.61×10^{14}	0.39	0.74	1.94×10^{1}
Decommissioning mining	4.59×10^{11}	0.008	—	2.00×10^{-2}

Table 11.
Summary of radiation environmental impact of different types of facilities.

3.6 Estimation of effective doses for the public based on environmental monitoring data

The estimation is based on the actual situation of public exposure caused by the cumulative discharge of radionuclides in various facilities after the operation of uranium mining and milling facilities for many years. The environmental monitoring data represent the cumulative results of the operation of the facilities over the years. Measuring the public exposure dose is difficult. Therefore, this project uses environmental monitoring data to calculate the cumulative maximum individual effective dose for the public.

3.6.1 Evaluation procedure

The cumulative evaluation involves two key data, namely, environmental monitoring and environmental background. They are important for the determination of the critical resident groups based on an effluent predictive evaluation. The evaluation procedure is as follows: source term determination → effluent monitoring → evaluation based on effluent data → identification of critical resident group → determining the environmental monitoring object and the pathway of irradiation → environmental monitoring → environmental background value determination and deducting → evaluation based on environmental monitoring data.

3.6.2 Radiation environment background determination technology

Environmental background determination is a key technology of the dose evaluation based on monitoring data. It mainly solves whether the monitoring data is at the environmental background level and is the key technology to judge the subtracted background value.

Grubbs method is used to test the background environmental data before operation or the level of comparison points during operation and to eliminate suspicious data. The distribution test of background data is carried out by Shapiro Wilke (S-W method) and skewness and kurtosis method.

If the natural background data $X = \{X_1, X_2, \dots X_n\}$ obey the normal distribution, the arithmetic average value is usually used to characterize the environmental

Mine type	Pathway of irradiation	Collective Dose (person Sv/a)							
		Rn-222	U-238	U-234	Ra-226	Th-230	Po-210	Pb-210	total
Conventional mining in operation	Gas	3.67×10^{1}	4.09×10^{-2}	4.79×10^{-2}	4.79×10^{-2}	2.17×10^{-1}	2.93×10^{-3}	1.49×10^{-2}	3.70×10^{1}
	Liquid	—	2.49×10^{-3}	7.19×10^{-4}	3.83×10^{-3}	1.74×10^{-4}	4.46×10^{-3}	6.31×10^{-2}	7.48×10^{-2}
	Total	3.67×10^{1}	4.34×10^{-2}	4.86×10^{-2}	5.17×10^{-2}	2.17×10^{-1}	7.39×10^{-3}	7.80×10^{-2}	3.71×10^{1}
In situ leaching mining	Gas	6.45×10^{-1}	—	—	—	—	—	—	6.45×10^{-1}
	Liquid	—	—	—	—	—	—	—	—
	Total	6.45×10^{-1}	—	—	—	—	—	—	6.45×10^{-1}
Shut down mining	Gas	1.94×10^{1}	—	—	—	—	—	—	1.94×10^{1}
	Liquid	—	3.87×10^{-4}	4.15×10^{-4}	1.17×10^{-2}	4.03×10^{-4}	2.48×10^{-3}	3.94×10^{-2}	5.48×10^{-2}
	Total	1.94×10^{1}	3.87×10^{-4}	4.15×10^{-4}	1.17×10^{-2}	4.03×10^{-4}	2.48×10^{-3}	3.94×10^{-2}	1.94×10^{1}
Decommissioning mining	Gas	2.00×10^{-2}	—	—	—	—	—	—	2.00×10^{-2}
	Liquid	—	2.04×10^{-7}	1.95×10^{-7}	3.49×10^{-8}	4.19×10^{-8}	4.02×10^{-14}	7.46×10^{-11}	4.76×10^{-7}
	Total	2.00×10^{-2}	2.04×10^{-7}	1.95×10^{-7}	3.49×10^{-8}	4.19×10^{-8}	4.02×10^{-14}	7.46×10^{-11}	2.00×10^{-2}
Total	Gas	5.67×10^{1}	4.09×10^{-2}	4.79×10^{-2}	4.79×10^{-2}	2.17×10^{-1}	2.93×10^{-3}	1.49×10^{-2}	5.71×10^{1}
	Liquid	—	2.88×10^{-3}	1.13×10^{-3}	1.56×10^{-2}	5.77×10^{-4}	6.94×10^{-3}	1.03×10^{-1}	1.30×10^{-1}
	Total	5.67×10^{1}	4.38×10^{-2}	4.90×10^{-2}	6.35×10^{-2}	2.18×10^{-1}	9.87×10^{-3}	1.17×10^{-1}	5.72×10^{1}

Table 12.
Collective dose distribution for the residents within 5 km.

background value. When the test object does not meet the normal distribution, it is converted logarithmically to check, whether it is conform to a logarithmic or a normal distribution. If it conforms to the normal distribution after the test, the geometric mean value can be used to characterize the environmental background value. After testing, when natural background data cannot be converted to normal distribution by scale, or when the scale transformation is very complicated, natural background data is a skewed distribution, and then, the natural background data is usually characterized by the median and average difference of the sample data. The median represents the location of the median value of the sample data, which is not affected by the minimum and maximum of the sample data. It is used when there is a lack of sensitivity. The median is representative for all persons and recommended, when the sample size is small ($n < 50$).

Mine type	Conventional mining in operation			*In situ* leaching mining			Shutdown mining							Decommissioning mining
Mine name	**M Mine**	**J Mine**	**L Mine**	**T Mine**	**X1 Mine**	**X2 mine**	**B Mine**	**Q Mine**	**DC Mine**	**JH Mine**	**JR-1 Mine**	**JR-2 Mine**	**TC Mine**	**LC Mine**
Maximum possible public individual effective dose (mSv/a)	0.40	0.72	0.59	0.12	0.05	0.03	0.44	0.34	0.51	0.03	0.36	0.45	0.29	0.00
The inhalation pathway contributes to the (%)	100	25.8	100	100	98.4	—	100	100	100	100	100	100	100	—
The feeding pathway contributes (%)	—	74.2	—	—	1.6	100	—	—	—	—	—	—	—	—
The concentration contribution of radon (Bq/m^3)	18.8	15.4	66.4	12.9	2.3	—	20.6	18.4	23.9	1.5	28.0	21.0	32.6	

Table 13.
Results of computational evaluated doses based on environmental monitoring data.

3.6.3 Cumulative doses

Calculated effective doses based on environmental monitoring data are generally higher than those based on effluent data, with large differences (see **Table 13**).

As **Table 13** shows, the maximum individual effective dose for the public, based on environmental monitoring data, is calculated mostly between 0.25 and 0.5 mSv/a. The maximum individual effective dose for the public, based on environmental monitoring data, is dominated by the ^{222}Rn inhalation contribution. J Mine Yangjia infant group is the largest individual dose for the public. The main contribution pathway is by ingestion (74.2%), dominated by ^{210}Po in rice (56.9% of the total dose).

3.7 Evaluation of normalized doses

According to the 30-year radiation environmental quality assessment of China's nuclear industry and the environmental status of uranium mining and milling, before 1986, the collective dose of radioactive effluent produced by uranium mining and milling in China was 19.3 person Sv/a, among which the dose of gaseous and liquid effluent was 14.7 and 4.59 Sv/a, respectively. The key radionuclides that cause the largest environmental public doses are radon and its daughters. The key pathway is by inhalation. The radon and its daughters contribution of the uranium mine accounts for about 90.9% of the total gas-liquid dose, and that of mills 91.8% (Pan et al., 1989) [13].

Due to the development of the wastewater treatment technology, the contribution of the liquid pathway to the public collective dose has decreased further in the last 10 years. The contribution of the radon release caused by underground mining is as high as 97.29%, and other pathways are negligible. Therefore, the public effective dose caused by underground mining can be replaced by the public effective dose caused by radon in the return air shaft of mines. Calculated collective doses for normalized natural uranium caused by each return air shaft are shown in **Table 14**. The weighted

Conventional underground mining							*In situ* blasting leaching mining		
B mine	Q mine	C mine	J mine	M mine	DC mine	Weighted mean	L mine	DB mine	Weighted mean
4.73	2.56	21.3	8.88	10.0	30.7	12.1	554	38.1	373

Table 14.
Normalized natural uranium public collective dose person·mSv/tU.

Year	Normalized radon release, TBq/GWa	Average normalized collective effective dose, person·Sv/GWa	*In situ* leaching percentage,%
1986 ~ 1995	284	8.09	1.0
1996 ~ 2005	258	7.29	10.0
2006 ~ 2010	224	6.15	20.8
2011 ~ 2014	181	4.04	37.0
2015 ~ 2019	115	3.51	50.0

Table 15.
Normalized radon release and the public collective effective dose from the uranium mines.

Year	Uranium mining and milling	Uranium purification	Uranium conversion	Uranium isotope separation	Nuclear fuel element manufacturing	Nuclear power plant operation	total	Uranium mining and milling ratio, %
2006 ~ 2010	6.15	1.61×10^{-2}	7.62×10^{-5}	5.61×10^{-5}	4.60×10^{-3}	1.61×10^{-1}	6.312	97.44
2011 ~ 2014	4.04	1.34×10^{-2}	1.33×10^{-4}	1.23×10^{-5}	2.44×10^{-3}	6.44×10^{-2}	4.120	98.05

Table 16.
Normalized public collective doses of nuclear power chains (person Sv/GWa).

average of conventional underground mining is 0.0121 person·Sv/tU, and the weighted average of *in situ* blasting underground mining is 0.373 person·Sv/tU. In addition, the normalized collective effective dose of the public caused by *in situ* leaching mine was 2.97×10^{-4} person·Sv/tU. According to the proportion of conventional underground mining, *in situ* blasting leaching underground mining, and *in situ* leaching mining each year, the annual average normalized collective effective dose is estimated (see **Table 15**).

The normalized collective effective dose of the public caused by the Chinese uranium mining and milling production was 21 ~ 33 times higher than the 5-year average value reported in UNSCEAR 2008 [11]. Due to the popularization and utilization of the *in situ* leaching process, the normalized collective effective dose of uranium production in China has shown a decreasing trend (**Table 15**).

In the nuclear power chain, the collective effective dose caused by mining and milling accounts for more than 97%, as shown in **Table 16**. The collective effective dose of the public caused by uranium mining and milling has increased from 80% after the establishment of the nuclear industry to 97%. It is mainly because uranium mining and milling is an open environment and the uranium deposits in China are small and scattered, and a large number of uranium mines have been developed in the past 30 years.

Evaluation period	Source term type	Release amount (Bq/a)								Ratio (%)
		Rn-222	U-238	U-234	Ra-226	Th-230	Po-210	Pb-210	total	
Current stage evaluation	Gas	7.42×10^{14}	8.94×10^{10}	4.60×10^{8}	4.38×10^{8}	4.60×10^{8}	4.38×10^{8}	4.38×10^{8}	7.42×10^{14}	99.9
	Liquid	—	6.89×10^{9}	6.63×10^{9}	1.49×10^{9}	1.15×10^{9}	1.21×10^{9}	1.95×10^{9}	1.93×10^{10}	0.1
	Total	7.42×10^{14}	9.63×10^{10}	7.09×10^{9}	1.92×10^{9}	1.61×10^{9}	1.65×10^{9}	2.39×10^{9}	7.42×10^{14}	100
	Ratio (%)	100	1.30×10^{-4}	9.56×10^{-4}	2.59×10^{-4}	2.17×10^{-4}	2.22×10^{-4}	3.22×10^{-4}	100	—
Previous 30 years of evaluation of the nuclear industry	Gas	2.76×10^{14}	1.79×10^{9}	1.79×10^{9}	3.94×10^{8}	3.68×10^{8}	3.72×10^{8}	3.68×10^{8}	2.76×10^{14}	99.9
	Liquid	—	9.55×10^{10}	9.55×10^{10}	7.09×10^{10}	1.22×10^{10}	1.01×10^{10}	1.85×10^{10}	3.03×10^{14}	0.1
	Total	2.76×10^{14}	9.73×10^{10}	9.73×10^{10}	7.13×10^{10}	1.26×10^{10}	1.05×10^{10}	1.89×10^{10}	2.76×10^{14}	100
	Ratio (%)	99.9	3.5×10^{-2}	3.5×10^{-2}	2.6×10^{-2}	4.5×10^{-3}	3.8×10^{-3}	6.8×10^{-3}	100	—

Table 17.
Comparison of gaseous and liquid radioactive effluent releases from uranium mines.

Compared with the evaluation of the first 30 years of China nuclear industry, there are great differences in facilities or mine objects, production scale, monitoring and evaluation methods, and data statistics types. Therefore, a comparative analysis was performed only from the overall conclusions. The comparison of source release is shown in **Table 17**, and the comparison of individual effective dose of critical resident groups is shown in **Table 18**. The comparison of collective doses caused by each nuclide is shown in **Table 19**.

Compared with the previous 30-year evaluation period of the nuclear industry, the total release of gaseous and liquid radioactive effluent from the uranium mining and milling facilities has increased by a factor of about 2.7. The contribution of gaseous effluent and ^{222}Rn remained basically unchanged, above 99.9%, due to the growing development of Chinese nuclear industry and increasing trend of the production scale. Due to the increased proportion of *in situ* leaching facilities and shutdown stage of underground mining facilities, the total amount of radionuclide release from the uranium mining and milling facilities increases, public maximum individual effective dose of the surrounding critical residential group is reduced, and in particular, the group part with less than 0.1 mSv/a has increased. The public collective dose of uranium mining and milling facilities has increased by about two times compared with the previous 30-year evaluation period of the nuclear industry. This is related to increased emissions of source terms (see **Table 17**) and increased population around

Evaluation period	**Maximum individual effective dose in critical resident groups (mSv/a) ***				
	<0.1	**≥0.1, < 0.25**	**≥0.25, < 0.5**	**≥0.5, < 1**	**≥1**
Current stage evaluation	64.3	7.1	21.4	7.1	0
Previous 30 years of evaluation of the nuclear industry	37.3	22.0	33.8		6.9

**Results based on effluent calculation.*

Table 18.
Proportion of individual effective dose distribution from uranium mines (%).

Evaluation period	**Pathway of Irradiation**	**Collective Dose (person Sv/a)**								**Ratio (%)**
		Rn-222	**U-238**	**U-234**	**Ra-226**	**Th-230**	**Po-210**	**Pb-210**	**total**	
Current stage Evaluation	Gas	56.7	0.041	0.048	0.048	0.217	0.003	0.015	57.1	99.8
	Liquid	—	0.003	0.001	0.016	0.001	0.007	0.103	0.130	0.2
	Total	56.7	0.044	0.049	0.064	0.218	0.010	0.118	57.2	100
	Ratio (%)	99.1	0.077	0.086	0.111	0.381	0.017	0.205	100	—
Previous 30 years of evaluation of the nuclear industry	Gas	12.70	0.885	0.055	0.026	—	0.013	0.115	13.80	76.7
	Liquid	—	0.233	0.014	0.948	—	0.055	2.94	4.19	23.3
	total	12.70	1.12	0.069	0.974	0.00	0.068	3.06	18.0	100
	Ratio (%)	70.6	6.2	0.4	5.4	0	0.4	17.0	100	—

Table 19.
Comparison of collective dose caused by each nuclide of effluent from uranium mines.

the facilities. In terms of the exposure pathway, the contribution of the liquid exposure pathway decreased by 23.2% and the gaseous exposure pathway increase by 28.5% due to the closure of facilities and the application of *in situ* leaching process and increasing trend of production scale.

4. Conclusions

This study develops the monitoring device for the determination of the radon release in the return air shaft of mines and the monitoring technology, which reveals release, migration, and diffusion behavior of radon and aerosols inside and outside the return air shaft. The method to monitor the radon release of liquid transport pipes and evaporation pools of *in situ* leaching mines, the radon release, and the monitoring method of radon in each vent of a uranium milling plant of *in situ* leaching are also established. The law of release, migration, and diffusion of radon and the distribution characteristics of gas-borne pollutants of *in situ* leaching mines are analyzed and determined.

^{222}Rn is still the dominant radionuclide, which is released from uranium mining and milling sources (more than 99.9% of the total gas-and-liquid pathway). The normalized radon release amount of various uranium mining and milling facilities types in decreasing order is operational *in situ* blasting leaching mine, operational underground mine, and operational *in situ* leaching mine, of which *in situ* leaching appears obvious environmental advantages.

The radon activity concentration at the outlet of the return air shaft of a mine shows an inverse parabolic function of its activity profile. Below the return air shaft head (-0.5 m, -1.0 m) a wavy line pattern of the activity profile was measured. The maximum radon activity concentration around the return air shaft was in the area of 100- to 300-m distance downwind. The maximum radon activity concentration around *in situ* leaching facilities occurs within a radius of 100 m. The radon activity concentration outside a distance of 500 m drops to the background level. In the gaseous effluent, the activity concentration of ^{210}Pb is several times higher than that of ^{238}U and 1–2 orders of magnitude higher than ^{226}Ra. The source term data are 1–2 orders of magnitude higher than for nearby residents. The activities of radionuclides in aerosols from source terms and in aerosols from the environment are ranged generally in: ^{210}Pb > ^{210}Po > ^{238}U > ^{226}Ra. In a few cases, ^{238}U > ^{210}Po. Radionuclide concentrations in aerosols from mines are 6 to 10 orders of magnitude lower than the corresponding radon activity concentration. Radionuclide concentrations in environmental aerosols are 4–6 orders of magnitude lower than the corresponding radon activity concentration.

The public effective doses caused by uranium mining/milling facilities are in descending order: *in situ* leaching operational mines > non-*in situ* leaching closed mines > *in situ* leaching operational mines > decommissioning mines. Over the past 30 years, the normalized public collective effective dose was reduced yearly, from 8.09 Sv/Gwa (during 1986–1995) to 3.51 Sv/Gwa (2015–2019). The normalized public collective effective dose is only 0.063 person Sv/GWa for residents around in *in situ* leaching facilities. The contribution of gaseous effluent and ^{222}Rn to the collective effective dose of the public increased from 70% to more than 99%. And the contribution of the liquid pathway to the collective dose was reduced from 23.3 to 0.2% due to the application of the *in situ* leaching uranium technique, closure and decommissioning of some facilities, and an effective management of the wastewater from mines. *In situ* leaching uranium mines show environmental advantages.

In short, this study provides methods and technical support for the daily operation at China's uranium mines and mills, such as the management of effluents, the environmental monitoring, pollution prevention and control, decommissioning, and environmental governance. It improves the radiation management of uranium mining and milling. It also provides important guidance for a pollution prevention and control and the supervision of the development of associated radioactive mineral resources. It has promoted the technological development of related industries and fields.

Author details

Lechang Xu*, Hui Zhang, Yalan Wang, Jie Niu and Xiangnan Dai
Beijing Research Institute of Chemical Engineering and Metallurgy, CNNC, Beijing City, China

*Address all correspondence to: xu_lechang@163.com

References

[1] El-Fawal Mohamed Mahmoud. Mathematical modeling for radon prediction and ventilation air cleaning system requirements in underground mines. Journal of American Science. 2011;7(2):389-402

[2] Patitapaban S, Panigrahi DC, Mishra DP. A comprehensive review on sources of radon and factors affecting radon activity concentration in underground uranium mines. Environmental Earth Sciences. 2016; **75**(7):1-19

[3] Huijia L, Feng X. Method for continuous monitoring of radon emissions from exhaust outlet in uranium mines. China Resources Comprehensive Utilization. 2020; **38**(4):1-4

[4] Mudd GM. Radon releases from Australian uranium mining and milling projects: Assessing the UNSCEAR approach. Journal of Environmental Radioactivity. 2008; **99**(2):288-315

[5] Dong X. Numerical Simulation and Experiment Investigation on the Radionuclide of Uranic Well Ventilation Exhaust Gas in Atmosphere. China: Central South University; 2009

[6] Wenxian W, Zhang Xiaowen X, Lechang.et al. Application of MATLAB for radon diffusion simulation in exhaust ventilation of uranium mine. Uranium Mining and Metallurgy. 2014;**33**(3): 147-150

[7] Yongjun Y, Shi L, Chunhua H, et al. Numerical simulation of radon radioactive effects for the surface around uranium mine exhaust outlet. Journal of University of South China (Science and Technology). 2018;**32**(1):1-9

[8] Xiaojun L. Characteristics of Radon Exhalation and Analyses of Radiation Dose from some in-Situ Leaching Uranium Mine of Xinjiang. China: University of South China; 2016

[9] Hui Z, Jie N, Deng Jun X, Lechang. Estimating the discharge amount of radon from production wells of CO_2+O_2 uranium mine. Journal of Isotopes. 2023; **36**(3):320-328

[10] Ziqiang P, Zhuzhou C, Zhibo W, Jianlun X. The radiological environment impact in China nuclear industry over past 30 years. Radiation Protection. 1989;**9**(4):241-247

[11] UNSCEAR. Report: Sources, Effects and Risks of Ionizing Radiation. New York, NY: United Nations, United Nations Scientific Committee on the Effects of Atomic Radiation; 2000. Available from: http://www.unscear.org/unscear/en/publications/2000_1.html

[12] Cimorelli AJ, Perry SG, Venkatram A, et al. AERMOD: A dispersion model for industrial source applications. Part I: General model formulation and boundary layer characterization. Journal of Applied Meteorology and Climatology. 2005; **44**(5):682-693

[13] Ziqiang P. The Radiological Environment Impact Assessment of China Nuclear Industry over Past 30 Years. Beijing, China: China Atomic Energy Press; 1990. pp. 89-99

Chapter 6

Radon Risks Evaluation and Application in the Assessment of Tremor-Prone Areas

Abdullahi Suleiman Arabi, Zainab Tukur,
Idris Isa Funtua, Musa Abdullahi Ali, Ewa Kurowska,
Musa Suleiman Abdulhamid and Adam Suleiman Murtala

Abstract

In just a time span of 5 years, northcentral Nigeria has experienced about three earth tremors. Public commentators were quick to attribute the tremors to the over-abstraction of groundwater and construction activities in the area. This study sets out to investigate the cause of tremors in the study area and develop a scientifically viable method that predicts areas prone to earth movement in the event of future tremors using integrated techniques (geophysical, hydrogeological, nuclear, and GIS). The results obtained confirmed the incidences of tremors in parts of northcentral Nigeria. Field and other evidence obtained suggest that the tremors were caused mainly by quarry activities in parts of the area. This suggestion was corroborated by results obtained from the surface manifestation of movements, radon in the groundwater, quarry proximity, slope, etc. Also, the study discovered substantial radiological implications associated with radon in groundwater because of earth movement. It also confirmed as baseless the notion of associating the tremors in the area with the over-exploitation of groundwater. Finally, the study came up with a method that predicts areas prone to earth movement in the event of future tremors in the area and categorized them based on their likelihood of experiencing movement as highly likely, likely, and not likely.

Keywords: tremor prediction, Northcentral Nigeria, radon, earth movement, seismic

1. Introduction

Despite the general belief that Nigeria is not situated on any recognized active seismic belt, quite a few earth tremors have been recorded since 1933 [1]. These tremors suggest that Nigeria may not be as aseismic as earlier envisioned. The intensities of these seismic events varied from 3.0 to 6.0, according to Akpan and Yakubu, 2010. These earthquake events were not given the needed attention by relevant authorities until lately. According to the National Space Research and Development Agency (NASRDA), one of the tremor events in northcentral Nigeria, which happened on Friday, September 7, 2018, at 5:11:32 am, had a moment magnitude of 2.6

IntechOpen

and a local magnitude of 3.0. These seismic activities have prompted renewed calls for proper investigations of earthquake possibility, particularly in northcentral Nigeria. However, conflicting attempts to reveal their source have led to assumptions on whether these earth tremors are tectonic or non-tectonic related [2]. Therefore, there is a need for scientific-based evidence to explain the origin and potential impact of these earth tremors. Shiwua [3] suggested that the incidence of earth tremors should not be overlooked because frequent tremors could be a build-up and forewarning of a major earthquake. Moreover, the lack of intraplate earthquakes makes them unpredictable, resulting in adverse consequences. Even with the occurrence of these events in Nigeria, practical measures such as the establishment of a database on groundwater quality, quantity, dynamics, vulnerability, and structural reinforcements for slope stability and to break floods, for example, are not well thought out.

According to Adepelumi et al. [4], a faulting event that resulted in previous earth tremors experienced in northcentral Nigeria was probably provoked by massive quarrying and blasting activities in the area. Despite the stress analysis conducted by the study, this concept is overly simplistic. Oluyide and Okunola [5] attributed the tremors in north-central Nigeria to non-tectonic slope instability and avalanches.

This study aims to develop a method that has the potential of predicting areas prone to movement in the event of future tremors in the study area. It seeks to reiterate why it is imperative to integrate the obvious knowledge of earth's dynamism and Nigeria's seismicity history. Achieving this may lead to measures that might aid the confining subsurface tectonic activities, limit potential damages resulting from earthquakes, and considerably reduce the potential risks to humans.

A unique aspect of this work is the incorporation of radon gas as a parameter in the prediction of earth movement, which ordinarily is not the case in most work of this nature. Natural disasters have a great influence on the national economy and social development. The precursors and prediction unavoidably remind one of the times when people tried to relate previous observable phenomena to an earthquake. Earlier observations were about changes in the water level in wells, distinct spring flows, and the burning and cracking of the land. Today, it may be difficult to observe these phenomena; however, they may be quantified, and other parameters that cannot be detected without instruments are monitored.

Radon is a natural gas that is produced in geogenic materials through the radioactive decay of the radium element. Radon can leave rocks and soils more easily by escaping into fractures and pore spaces between grains of soil. Measurements of this gas both in soil and in groundwater have shown that spatial and temporal variations can provide knowledge about geodynamic events. Radon has potential use in earthquake prediction studies because its concentration variation in the subsurface has been observed to precede an earthquake. The first evidence of a correlation between radon and earthquake occurrence came from an observation of radon concentration in well water prior to the Tashkent earthquake of 1966 [6]. This evidence stimulated research work in this area soon afterward in many countries. Radon observations, both in soil gas and in groundwater, revealed many premonitory changes in its concentration before an earthquake [7–16]. As a result of these experimental correlations, radon is considered one of the few promising signs that can be used in earthquake prediction studies.

The first measurements of radon as a seismic precursor dated back to 1927, but the first recording that encouraged research on seismic precursors was detected before the Tashkent earthquake of 1966. The pre-earthquake stress gradients mobilized underground fluids, carrying some dissolved gases to the near-surface region, mainly through faults, lineaments, and thermal areas [16]. This kind of anomaly may sometimes last for a short duration; therefore, continuous monitoring of sub-soil radon is necessary to correlate such anomaly with earthquake.

To set up an automatic radon monitoring station, geologically sensitive locations are selected for the measurement of subsoil radon concentrations. This suggests a relationship between the investigated indicators and tectonically related processes related to the tremors. These signs are further classified into: (a) radon exhalation from the earth's crust; (b) exhalation of helium, argon, and others; (c) temporal variation in water level or discharge of springs; and (d) temporal variation in temperature and dissolved ions in the water of the monitoring sites [17].

2. The study area

The area covered by this work is located within latitude 9° 03′ 28.26” N and longitude 7° 29′ 42.29″ E. It covers the entire Federal Capital Territory (FCT) (**Figure 1**). The widespread rock units of this area consist of “Older granites” of the pan-African granitoid and migmatite-gneiss complex. This comprises an inhomogeneous cluster of gneisses (ortho- and para-), migmatites, and successions of basic and ultrabasic rocks. Petrographically, the pan-African Orogeny caused the recrystallization (via partial melting) of the migmatite and gneissic rocks.

The migmatite-gneiss complex comprises rocks that vary between 600 million years and 3 billion years, that is, from pan-African to Eburnean, which is reflected by the numerous chrono-successions of the migmatite-gneiss complex, the majority of which originated during the orogenic cycle. The rocks of the migmatite-gneiss

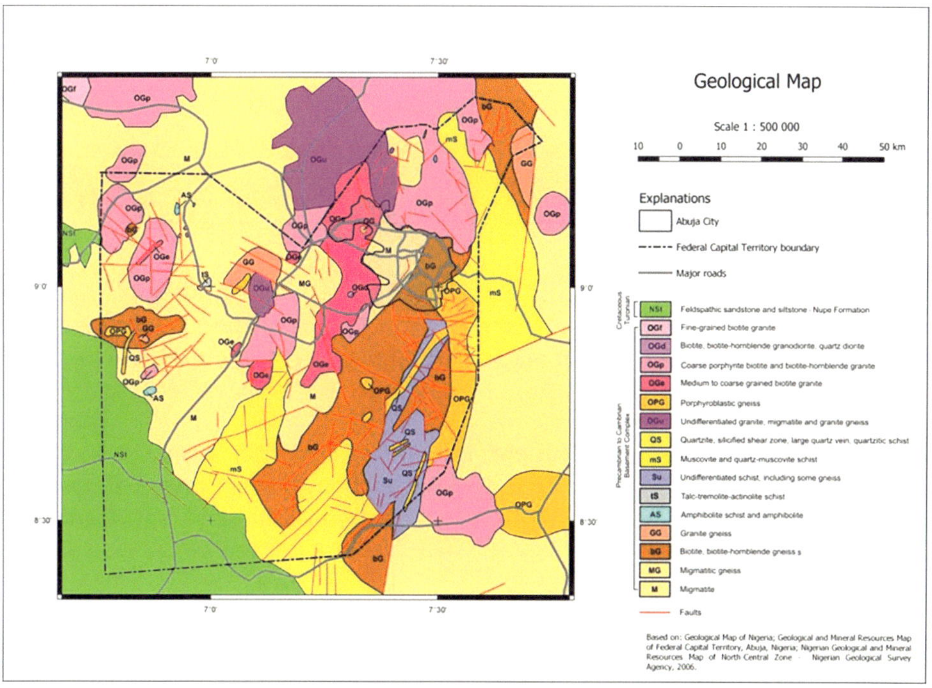

Figure 1.
Updated geologic map of the FCT based on field investigation and complemented by [18] (geologic and mineral resources map of the FCT) (inset: Map of Nigeria depicting the location of the study area).

complex are characterized by varying mineralogy, chemical composition, texture, and structure. These variations are attributable to the progressive/gradual regional metamorphism of meta-sedimentary rocks [19].

3. Research method

The methods utilized in this study are divided into: (a) geological, (b) hydrogeological, (c) geophysical, (d) nuclear, and (e) geographic information systems (GIS). These methods are discussed further based on the objectives of the study as follows:

a. Detailed field studies of structural patterns and geology of parts of northcentral Nigeria affected by tremors.

b. Geophysical study of the area to ascertain the different subsurface geologic units and groundwater saturation of each unit.

c. Evaluation of the trend of earth movement and its effect on groundwater quality, quantity, vulnerability, and dynamics in the area.

d. Determination of the cause of sudden movement within the subsurface of the area.

e. Evaluation of the safety or otherwise of activities such as borehole drilling, quarrying activities, sand mining, and construction activities on earth movement in the area.

The study area was gridded, and measurement was taken for each of the methods (geophysical, hydrogeology, nuclear, and geological investigations) aimed to achieve the following:

a. Geological investigations: This involves the study of geological structures and their trend, lineaments, and rock types in the study area (depicted in **Figure 2**).

b. Hydrogeological investigations: This entails the study of hydrochemical/physicochemical parameters of groundwater, water table, flow pattern, aquifer types and production of hydrogeological map, and compliance of water quality to national and international standards.

c. Geophysical investigations: This entails the study of gravity maps, electrical resistivity profiling, geo-electric section production, production of basement relief map, slope and water percolation investigation, and earth movement prediction.

d. Nuclear technique: The technique involves the measurement of radon in groundwater and the evaluation of its seismic relationship and radiological implication on the health of the inhabitants of the area, radon exhalation and emanation coefficient in rocks, evaluation of seismic relationships, determination of the structure regarded as radon entry points, and the radiological implication of air quality in the area.

Figure 2.
Photograph of the (a) geology team depicting some activities, (b) hydrogeology team taking the measurement of water quality parameters, (c) geophysical team conducting vertical probing of subsurface geo-electric layers, and (d) nuclear team conducting radon measurement in groundwater sample and depth profiling.

e. Geographic information system (GIS): The technique employed to produce the digital elevation model (DEM), lineament extraction, superimposition of different maps for information deductions, and update of geologic and other maps of the area.

The techniques above were utilized to derive/deduce relevant information with regard to earth movement, faulting, radon entry points, groundwater quality and quantity, radiological implication, and prediction/forecasting of areas prone to movement in the case of future tremors.

3.1 Radon measurement in groundwater and subsoil

The setup for radon in the groundwater is designed to achieve the close loop concept of measuring radon in water as provided in the RAD 7 manual (**Figure 3a**). A detail of the protocol of radon measurement in groundwater and subsoil can be found in [19].

3.1.1 Radon in groundwater

As discussed above, radon has potential use in earthquake prediction studies because its concentration variation in the subsurface radon has been observed to

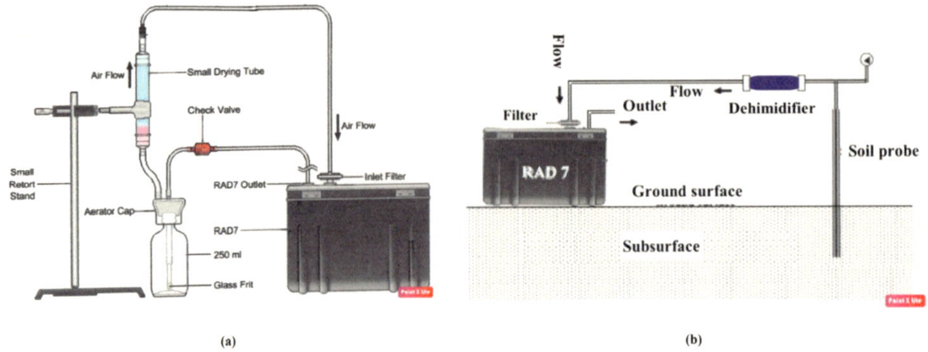

Figure 3.
(a) Setup adopted during in situ measurement of radon concentration in groundwater [20]; (b) setup for soil radon depth profiling.

precede an earthquake. The first evidence of a correlation between radon and earthquake occurrence came from observing radon concentration in groundwater before the Tashkent earthquake of 1966 [6]. Radon concentration in groundwater from different locations in the study area was used as one of the parameters for the predictive method and is given a weight of five (5).

3.1.2 Radon in the subsoil

Radon observations, both in soil gas and in groundwater, revealed many precursory changes in its concentration before an earthquake [7]. As a result of these experimental correlations, radon is considered one of the few promising precursors that can be used for earthquake prediction studies. Radon in the subsurface is also used as an input in the method developed and was given a weight of five (5).

3.1.3 Groundwater percolation

Percolation is the movement of water through the soil itself. Percolation contributes greatly to earth movement by initiating the movement of the mass of rock, debris, or earth down a slope under the direct influence of gravity. In this method, percolation was also used as one of the parameters and was given a weight of ten (10).

3.1.4 Elevation

The elevation is a factor that can be responsible for inducing slope instability. Higher elevation has been attributed to increased surface runoff. There is also evidence that suggests elevation has a major influence on the spatial and temporal distribution of the landslides. In this study, elevation was also considered as a parameter and was given a weight value of ten (10).

3.1.5 Quarry proximity

The work conducted has discovered that quarry activities in the study area are the main culprit implicated for earth movement in the tremor, which results in the tremors being felt. Also, it was discovered that surface manifestation of earth movements in the area is closely related to quarry proximity, meaning that all movement manifestations observed were proximally close to the quarries or far from the quarries. Therefore, quarry proximity was also used as a parameter in the prediction method and was given a weight value of ten (10) because it's a major contributor to earth movement.

3.1.6 Lineament

Lineaments are structurally controlled linear or curvilinear features identified by their relatively linear alignments. It is a linear feature in a landscape that is an expression of an underlying geological structure such as a fault. Lineament narrates in detail a much broader and more pertinent structural pertain that may be an indication of possible areas affected by movement within the subsurface. In the methods developed, lineament is one of the most important features showing subsurface elements or structural weaknesses such as faults and, as a result, is a parameter that is taken into consideration and assigned a weight value of five (5).

3.1.7 Slope/basement relief

The outward and downward displacement of slope-forming materials under the influence of gravity is greatly influenced by the slope. Once rock material is broken down into smaller, unstable pieces by weathering, the material has the potential to move down the slope. The steeper the slope, the greater the potential for gravity to pull earth material down a slope, and as a result, the slope is also a major parameter considered in the development of this method. In this case, the slope is assigned a weight value of ten (10). The consideration of basement relief as a parameter in the development of the prediction method is because it is a sloppy continuation in the subsurface. It has the tendency of influencing subsurface movement, especially when associated with percolation and groundwater abstraction.

The total weight value assigned to all the eight parameters considered in the prediction method is sixty (60), and the earth movement prediction map that could be generated will categorize an area as likely (weight between 40 and 60), moderately likely (between 20 and 39), and unlikely (between 0 and 29).

3.2 Validation of the method

The prediction method (RQ-PABLES METHOD) developed for the prediction of areas prone to earth movement was validated in the field using areas/locations with a surface manifestation of movement and quarry proximity. The method was developed using cumulative weight assigned and generated from seven parameters, namely, percolation, elevation, radon in groundwater/radon in sub-soil, lineament, quarry proximity, basement, and slope. The weight value ranges from 1 to 10 depending on the influence of each of the parameters that could result in triggering the earth movement.

4. Results and discussion

The occurrence of earth tremors contradicts the earlier thought of the aseismic nature of Nigeria. It is relevant to mention that some of the villages and towns troubled by the recent seismicity are located on the Atlantic fracture system. As suggested by Adepelumi et al. [4], the SW-NW trend of the fault makes that stretch of land vulnerable to seismicity because of stresses produced within the earth's crust, that is, partial reactivation of the ancient faults. Displacements and fractures were explored in different outcrops across several localities in Abuja possibly associated with the NNE–SSW faults, and their orientations were measured. A collection of the observed faults and fractures are presented in **Figure 4a**. The mapped fracture trends (**Figure 4a**) define a major NW-SE trajectory as shown on the rosette plots (**Figure 4b**). Logically, the measured trends might denote that the tremor is tectonic based on the regional geological background (the North Atlantic fracture systems), field measurements of the structural patterns, and prior aeromagnetic data. However, this rationale is subjective and hindered by the dearth of seismic data. Goki et al. [2] in their study related the recurring earth tremors to the reactivation of fault systems associated with these Atlantic Fracture Zones. This deduction is supported by the fact that the tremor epicenters of the recent Mpape tremor in Abuja fall along the major fractures and their projections appear to align with the Chain Fracture Zone in the Atlantic. However, the mechanism for the reactivation was not fully elaborated on or explained by Goki et al. [2]. Based on our observations carried out around quarry sites, reports from inhabitants in the vicinity insinuate that the earth's tremors can be linked to the current pervasive and uncoordinated detonating and blasting activities performed at quarry sites. Accordingly, the new fractures crosscutting the major faults of the gneissic basement complex were noted at the quarry sites as well as road cuts (**Figure 5**). Although it was not established if these new fractures were deep-seated, the discordance and fractures they created are evidence that supported the reactivation theory.

Most of the structures observed in the field are northwest–southeast oriented (**Figure 4a** and **b**) and are mainly found around areas with high quarrying activities such as Mpape and Dutsen-Alaji. These structures were neither found to be associated with certain rock types nor appear to be induced by human activities in those areas. An important field observation was the identification of some volcanic rocks (basalt and pumices) in the study area, which were not reported by earlier works in the

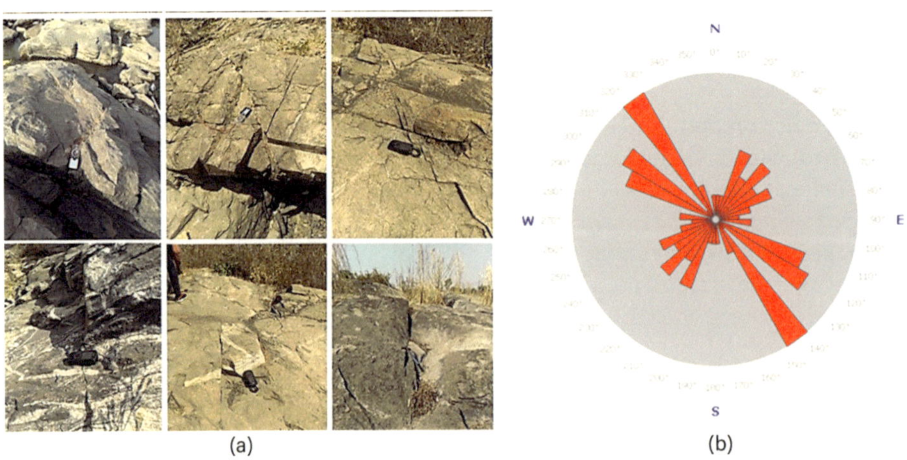

Figure 4.
(a) Fracture trends from field measurements on granitic rocks. (b) Rose diagram indicating the NW-SE trend of the fractures.

Figure 5.
Field photographs of (a) rock slump, (b) clusters of fragmented boulders at Mpappe, (c) major fracture at a road cut at Wasse, (d) boulders set at a hill at Bwari, (e) landslide behind a quarry site at Dutsen Alaji, and (e) fragmented granite boulder.

area. This development needs to be followed up to ascertain the extent and nature of occurrences.

4.1 Parameters of movement prediction

The parameters used in the prediction of locations prone to movement in the case of future tremors in the area are radon in groundwater, radon in the subsoil, percolation, elevation, quarry proximity, lineament, slope, and basement configuration. These parameters were assigned weighted values depending on their perceived contribution to earth movement as indicated earlier.

4.1.1 Radon in groundwater

A summary of the result of the measurement of radon in 128 samples of groundwater is provided in **Table 1**. These include the activity concentration of radon in groundwater from the area, which ranges from 609.00 to 92,500.00 Bqm^{-3} (0.61 to 92.50 BqL^{-1}) with a mean value of 16628.19 Bqm^{-3} (16.63 BqL^{-1}). These values

	Radon in groundwater (Bq/m3)	Radon in groundwater (Bq/L)	Annual effective dose from ingestion (mSv/y)	Annual effective dose from inhalation (mSv/y)	Total annual effective dose (mSv/y)	TDS (mg/L)	TUB (NT)	COND (mS/cm)	TEMP (°C)	pH
Lowest	609	0.61	0.13	0.002	0.13	1.03	0.24	8.72	25.5	1.53
Highest	92500	92.5	19.43	0.23	19.66	548	429	1457	39.28	14
Mean	16628.19	16.63	3.49	0.04	3.53	127.46	19.81	194.32	32.1	6.76

TUB = turbidity, COND = conductivity, TEMP = temperature.

Table 1.
Highest, lowest, and mean values recorded for radon in the groundwater; calculated annual effective dose due to ingestion and inhalation; total annual effective dose; and some physicochemical parameters.

exceeded both the EPA's MCL of 11.1 BqL^{-1} and the NIS limit of 0.1 BqL^{-1}. The calculated annual effective dose due to ingestion and inhalation of groundwater from the area ranged from 0.13 to 19.43 $mSvy^{-1}$ with an average of 3.49 $mSvy^{-1}$ for ingestion, while the range was 0.002 to 0.23 $mSvy^{-1}$ for inhalation and a total annual effective dose with a range between 0.13 and 19.66 $mSvy^{-1}$ with an average of 3.53 $mSvy^{-1}$.

The dose coefficients given in ICRP Publication 137 were determined based on a dosimetry approach. This considers the physical phenomena that determine the distribution of radionuclides in the body to quantify the energy deposited per unit mass in the various regions of the body. It also gives a weighting based on the toxicity of the radiation and the radiosensitivity of the tissues using an epidemiological approach.

4.1.1.1 Radon in groundwater as a precursor to earth movement

The results of radon in groundwater suggest that radon in groundwater from the study area is closely related to other physicochemical parameters of the water such as temperature, and this might not be unconnected with the local geology of the area, geothermal gradient, depth to aquifers, and the possibility of radon entry point created by seismic activities experienced in the area in the past 5 years, which causes radon release from basement fractures into groundwater bodies. The tainting of groundwater by radon gas to the spatial extent (**Figure 6**) seen in these results might also suggest the widespread nature at which radon entry points are created within the surface. Movements within the subsurface potentially create openings (fracture) within the basement, which are referred to as "radon entry points" through which radon gas is exhaled into groundwater bodies in both primary and secondary fractures, which serve as an aquifer.

4.1.1.2 Radon in the subsoil as a precursor to earth movement

Radon profiling at different depths (0.33, 0.66, and 0.99 m) within the subsurface was conducted at 130 locations that covered the entire study area. Radon activity concentration in soil at a depth of 0.33 m (33 cm) ranged from 15.20 to 48,500.00 Bq/m^3 with an average value of 4257.47 Bq/m^3. At a depth of 0.66 m

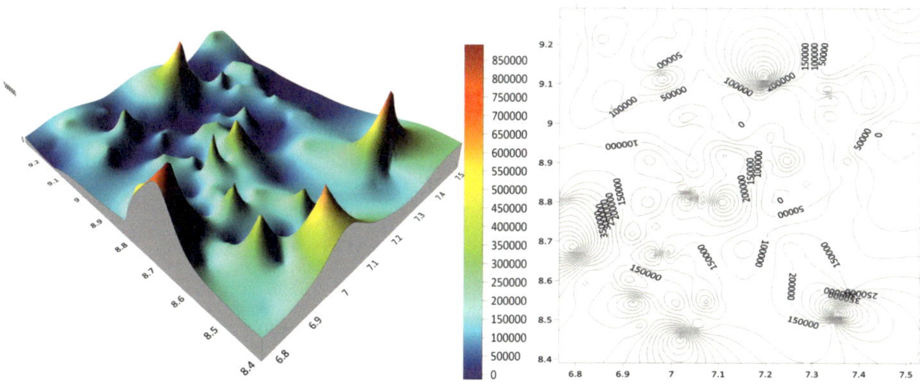

(a) (b)

Figure 6.
Radon activity concentration in groundwater: (a) 3D map; (b) contour map.

(66 cm) below the ground surface, the values obtained ranged from 15.20 to 59,600.00 Bq/m^3 with a mean of 5061.19 Bq/m^3, while at 0.99 m (99 cm) below the ground surface, the values recorded ranged from 10.02 to 81,200.00 Bq/m^3 with a mean value of 9993.15 Bq/m^3. About 88.5% of the values recorded were observed to have exceeded the 200 Bq/m^3 action level recommended by the International Commission on Radiological Protection (ICRP). High radon activity concentration coupled with the downward increase in radon within the subsurface may also be a pointer to the effect of seismic activities experienced in the study area during recent times. Earth tremors recently experienced in the area might have caused some openings (fractures) within the basement rocks, which eventually serve as radon entry points into groundwater and the overburden. Naturally, radon from the basement travel upward into the overburden, and this might manifest in the form of radon increase with depth. This result clearly indicates how radon activity concentration in the area increases with depth (**Table 2**).

4.1.2 Lineament

Lineament density depicts the frequencies of subsurface movement represented by subsurface structures such as fractures, veins, joints, and other indicators of points of weakness in the subsurface. Simply put, a lineament is a linear feature in a landscape that is an expression of an underlying geological structure such as a fault. Typically, a lineament appears as a fault-aligned valley, a series of fault or fold-aligned hills, a straight coastline, or indeed a combination of these features. The lineament map of the study (**Figure 7**) indicates higher lineament densities around the northeastern, southeastern, southwestern, and northwestern portions of the study area, while the central and western portions have the lowest density in that order. This suggests that the subsurface movement is likely to be intense in that same order. This also corresponds to the basement relief map (**Figure 7**), suggesting that movement is likely to be intense at depth. These are also areas where radon entry into groundwater is expected.

4.1.3 Basement relief and slope

The basement relief as extracted from vertical electrical sounding result indicates that the basement is the deepest in the northeastern part of the study area (**Figure 8**) and shallower in the central portions, while it is intermediate in the southwestern part. This means that the thicker overburden is in the northeastern part, while it is thinner

	Radon at 0.33m (Bq/m3)	Radon at 0.66m (Bq/m3)	Radon at 0.99m (Bq/m3)	In-situ radon diffusion length at 0.33 *(m)*	In-situ radon diffusion length at 0.66 *(m)*	In-situ radon diffusion length at 0.99 *(m)*
Min	15.2	15.2	10.02	0.005	0.005	0.003
Max	48,500.00	59,600.00	81,200.00	15.96	19.62	26.73
Ave	4,257.47	5,061.19	9,993.15	1.4	1.67	3.29

Table 2.
Values at depths below earth's surface.

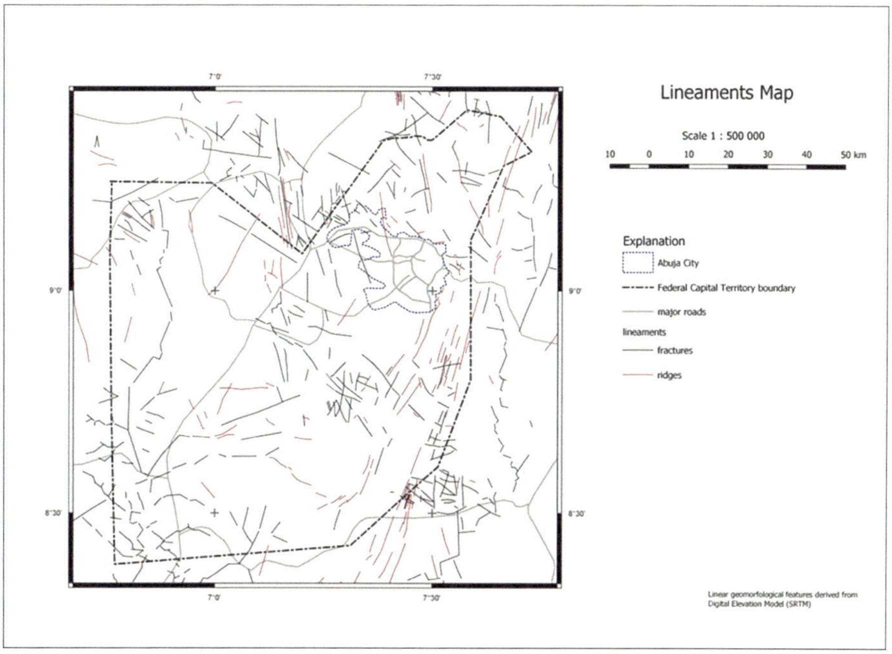

Figure 7.
Lineament density map of the study area.

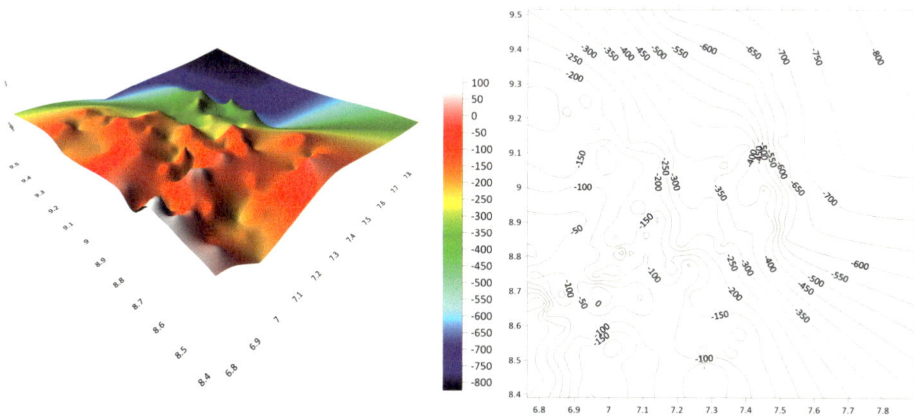

Figure 8.
3D basement relief/slope map of the study area.

in the central part of the study area. In terms of slope, which significantly has a control on earth movement with slight triggers, especially when percolation is high, even the central portion, where the basement is shallower, is an area with potential for earth movement/landslide.

4.1.4 Elevation

Elevation in the study area is the highest around the northeastern parts, while it is the lowest in the southwestern part. The north–south and northwestern portions are

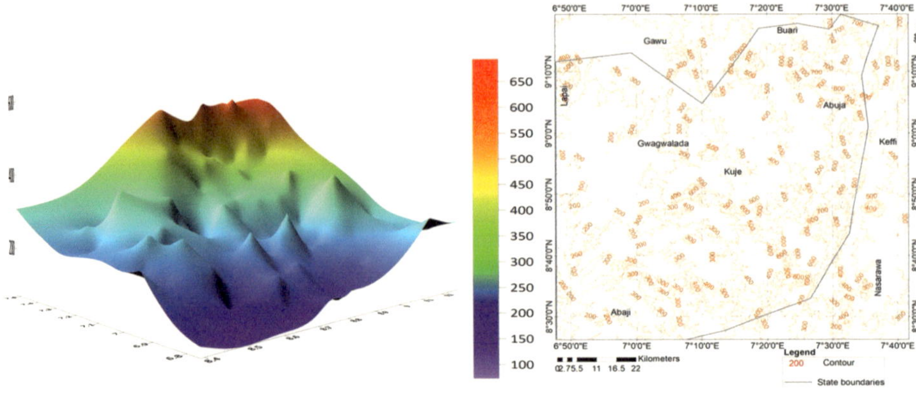

Figure 9.
3D elevation and contour map of the study area.

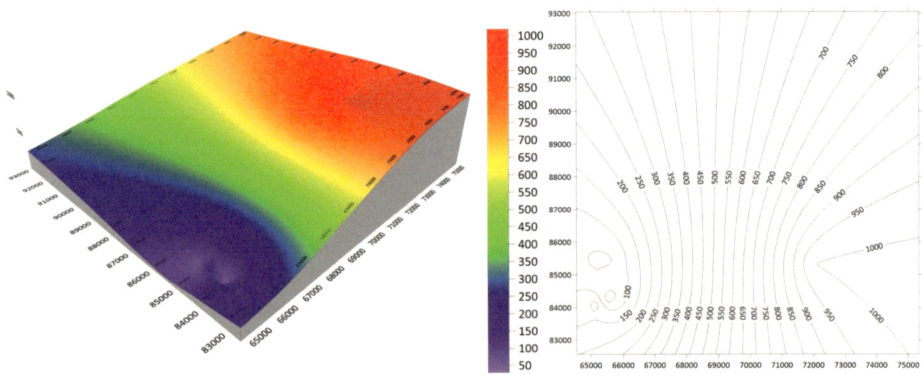

Figure 10.
3D quarry proximity and contour map of the study area.

relatively flattened, while the other parts are rugged and mountainous. Most of the quarries are situated in areas of high elevation, where outcrops are dominant (**Figure 9**).

4.1.5 Quarry proximity

The quarry sites in the study area are mostly located in the northeastern and southeastern parts. A plot of quarry proximity to a particular location gives an insight into the possibility of a quarry activity having a significant effect on the stability of that area. This result suggests that this effect is intense in the northeastern–southeastern parts, while it is the lowest in the western parts (**Figure 10**).

4.1.6 Groundwater percolation

The groundwater saturation within the subsurface for a particular area can influence the stability of the area, especially when the effect of slope in such an area is intense. Ordinarily, the movement of the earth material on the sloping terrain is down a gradient. Therefore, in areas saturated with groundwater, especially

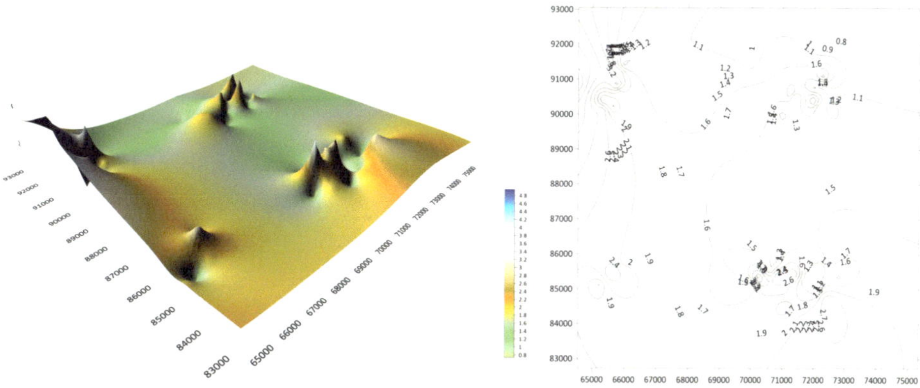

Figure 11.
3D and groundwater percolation contour map of the study area.

when the earth material around those areas is made of clays, the slightest trigger or agitation can cause the earth material to move downslope. In this study, areas identified as vulnerable to movement because of groundwater saturation are the northwest, southwest, northeast, and around the southwestern portion of the study area (**Figure 11**).

4.2 Earth movement prediction

The earth movement prediction map was produced from the cumulative weight of all weighted averages extracted from radon, lineation, quarry proximity, slope, elevation, and groundwater percolation maps. These maps were generated from the measurements made in situ during the field campaign. The superimposition of all the maps generated can give insights into the likelihood of areas that could experience movement in the event of future earthquakes or earth tremors in the north-central part of Nigeria (**Figure 11**). These areas were categorized based on their likelihood of experiencing movement as highly likely, likely, and not likely (**Figure 12a** and **b**).

4.2.1 Highly likely areas

The areas categorized as highly likely to experience earth movement in the event of a tremor or earthquake in the study area are mostly in the western parts of the study area. These areas also fall within locations where the surface manifestation of earth movement was observed during the field campaign. These areas include Mpape, Saupe, and Dushepe, all around the northeastern parts of the study area. These highly likely areas are further categorized into upper and lower or most highly likely, which are found around certain pockets of locations in the western parts of the study area around Burum, Sherete, and Gwau and highly likely (found close to the western edge and stretches/trends north southernly) (**Figure 12a** and **b**). Also, within the most likely areas are locations around Bogu, Tagwai, Kumeo, and Kashimoro, all around the southwestern parts of the area.

The relocation of quarry sites close to residential areas around this location is highly recommended to safeguard and protect likely damages in all ramifications in the area in the event of a future tremor.

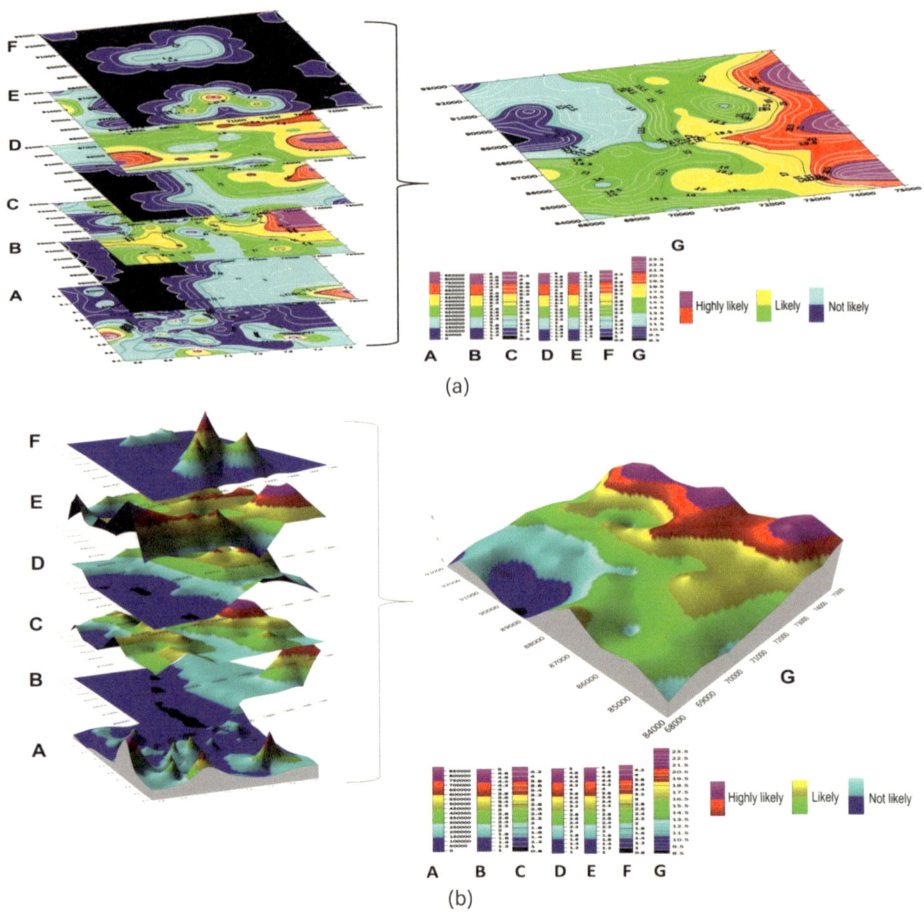

Figure 12.
(a) Superimposed 3D contour (b) surface maps of radon, quarry proximity, slope, elevation, lineament, and corresponding earth movement predictive map (A = radon in groundwater and subsoil, B = quarry proximity, C = slope, D = elevation, E = lineament, F = percolation, and G = predictive model).

4.2.2 Likely areas

The areas deduced as likely to experience movement in the event of future tremors are found in the north, south, and southwestern parts of the study area (**Figure 12a** and **b**) around Dakwa, Gwi, Gao, Yebu, Yabo, Nuku, Ebagi, and Abaji. These categories constitute/cover almost 50% of the study area. They are further categorized as most likely and likely. Most likely areas constitute about 30% of the total likely areas and are mostly found in the south, central, and northeastern parts of the area, while the likely areas constitute about 70% and are mostly located in the north, central, and southwestern parts of the study area. Surface manifestation of movement was observed neither in these areas nor in operational quarry sites. These areas are much safer than the areas discussed above; therefore, for the sake of the future, to obtain a license for quarry activities in these areas, there is the need to conduct and incorporate an earth movement prediction study in all environmental impact assessment reports. This is to make sure that the inhabitants are protected from any force majeure that could occur because of quarry activities.

4.2.3 Not likely areas

Areas categorized as "Not likely" are restricted to the northwestern parts of the study area (**Figure 12a** and **b**). Based on the assessment in this study, these areas are the safest; they are further categorized as "most unlikely and safe". The safe areas are on the western and northwestern corners. The not-likely areas cover areas around Shimbo, Pate, Dako, Ghamgwa, and Babbantsauni villages. It was observed that areas with many quarries operating at the same time are more susceptible to stress build-up, which eventually triggers sudden earth movement. This phenomenon is likely the cause of tremors of higher intensities with resultant surface manifestations of movement observed around Mpape and its environment.

4.3 Effect of other activities on earth movement

The records of tremors in the study area have prompted so many speculations, which have not been scientifically proven as discussed earlier in the introduction. Some of the speculations are that the tremor was a result of the over-exploitation of groundwater in the area, while others tie it to massive construction activities in the area. All these speculations have not been proven scientifically.

4.3.1 Groundwater over abstraction

From the results obtained, this study has not discovered any effect of groundwater abstraction on tremors that were experienced in the area. The study also did not find any association between groundwater quality, as it relates to the tremor records in the area. With regard to quarry activities, this might have implications on groundwater dynamics in the area in two ways as follows:

a. Fractures created by blasting during quarry activities can change the groundwater flow direction. This direction is controlled by hydrodynamics, hydraulic heads, and interconnectivity between fractures.

b. Groundwater quantity in the area might be affected both negatively and positively.

    i. Positively: As new fractures are created, the secondary porosity will be enhanced, serving as a groundwater storage conduit.

    ii. Negatively: The creation of fractures has the possibility of creating interconnectivity between fractures, which allows subsurface groundwater flow to change; as a result, groundwater levels from one location to another may change such that it decreases from one location and increases in another depending on the hydraulic characteristic of subsurface parameters related to groundwater.

    iii. These are possibilities that must be substantiated scientifically by monitoring certain parameters related to groundwater in the area over time.

4.3.2 Construction activities

Based on the finding of this study, apart from quarry activities, which may not be categorized under construction activities, other activities that are construction related do not result in tremors in the area as speculated by some commentators in the public domain. There is no doubt that the proliferation of quarry sites in the FCT has a direct implication on the resultant tremor that was felt many times around northcentral Nigeria, especially in the FCT.

5. Conclusion

This study set out to investigate reported cases of tremor in parts of northcentral Nigeria; study the causes and implications on groundwater quantity, quality, and dynamics; study the radiological implication of radon released into groundwater bodies because of the tremor; and study and come up with ways of determining areas prone to earth movement and mitigation measures to plan and contain future occurrence.

The study confirmed the incidences of tremors in parts of northcentral Nigeria. Field and other evidence obtained suggest that the tremors were caused mainly by the quarry activities in parts of the area. This suggestion was corroborated by results obtained from the surface manifestation of movements, radon in the groundwater, quarry proximity, slope, etc. Also, the study discovered substantial radiological implications associated with radon in groundwater because of the earth movement. The study also confirmed as baseless the notion of associating the tremors in the area with the over-exploitation of groundwater.

Finally, the study came up with a method that predicts areas prone to earth movement in the event of future tremors in the area.

5.1 Recommendations

This study did not attempt to find out the reasons behind the siting of quarries even in residential areas or areas already inhabited. Perhaps the siting/location of quarries in the FCT and environs is of dual objects: to explore and exploit the construction aggregate for the massive construction activities in the FCT or to pave the way or flatten rocky areas for future developmental activities; neither of these can be achieved without its advantages and disadvantages. The results discussed in this report are far-reaching in ascertaining the cause of tremor in northcentral Nigeria and its implications on the populace. To monitor future occurrences, mitigate effects, and reduce the possibilities and implications of the future occurrence due to activities of the likely culprits identified, the following recommendations are made:

a. Revocation, relocation, and reassigning of quarry licenses issued in the identified vulnerable areas.

b. Review and recommend more environmentally friendly blasting methods and procedures for quarry operators.

c. Include study of land movement and landslide possibilities in Environmental Impact Assessment (EIA) reports

d. Recommend standard distance for quarries from residential and other inhabited areas.

e. Conduct continuous quarterly radon assessments in groundwater to track and monitor spikes that could suggest earth movement.

f. Continuous monitoring of groundwater levels around FCT and environs to track quality, quantity, and dynamics, especially with regard to quarry activities in the areas.

g. The siting of licensed quarry should be well spaced such that, even if they operate at the same time, the stress built up would not be related in such a way that could trigger earth movement.

Author details

Abdullahi Suleiman Arabi[1*], Zainab Tukur[2], Idris Isa Funtua[3], Musa Abdullahi Ali[1], Ewa Kurowska[4], Musa Suleiman Abdulhamid[5] and Adam Suleiman Murtala[6]

1 Department of Geology, Bayero University, Kano, Nigeria

2 Department of Biological Sciences, Bayero University, Kano, Nigeria

3 Center for Energy Research and Training, Ahmadu Bello University, Zaria, Nigeria

4 Department of Forest Engineering, Poznan University of Life Sciences, Poland

5 Department of Physics, Bayero University, Kano, Nigeria

6 Department of Geology, University of Jos-Nigeria, Nigeria

*Address all correspondence to: asabdullahi.geo@buk.edu.ng

References

[1] Tsalha MS, Lar UA, Yakubu TA, Kadiri UA, Duncan D. The review of the historical and recent seismic activity in Nigeria. IOSR Journal of Applied Geology and Geophysics (IOSR-JAGG). 2015;**3**(1):48-56

[2] Goki NG, Onwuka SA, Oleka AB. Preliminary geological evidence for multiple tremors in Kwoi, Central Nigeria. Geo-environmental Disasters. 2020;**7**:22. DOI: 10.1186/s40677-020-00156-w

[3] Shiwua AJ. Seismicity in Nigeria: The need for earthquake resistant structure. Zeszyty Naukowe Politechniki Częstochowskiej seria Budownictwo. 2013;**19**(1):171-178

[4] Adepelumi AA, Ako BD, Ajayi TR, Olorunfemi AO, Awoyemi MO, Falebita DE. Integrated geophysical mapping of the Ifewara Transcurrent fault system, Nigeria. Journal of African Earth Sciences. 2008;**52**(4-5):161-166

[5] Oluyide PO, Okunola OA. Hill-slope wash and rock avalanche in Nandu area, Gwantu District, Jema'a L. G. a. of Kaduna state. In: Onuoha KM, Ofodile ME, editors. Proceedings of the International Workshop on Natural and Man-made Hazards in Africa, Awka, Anambra State, Nigeria. Nigeria: Nigerian Mining and Geosciences Society; Jan. 31-Feb. 3, 1993

[6] Ulomov VI, Mavashev BZ. A precursor of a strong tectonic earthquake. Academy of Sciences of the USSR Earth Sciences Sections. 1967;**176**:9-11

[7] Talwani P, Moore WS, Chiang J. Radon anomalies and microearthquakes at Lake Jocassee, South Carolina. Journal of Geophysical Research. 1980;**5**(B6):3079-3088. DOI: 10.1029/JB085iB06p03079

[8] Mogro-Campero RL, Fleischer RS. Likes Changes in subsurface radon concentration associated with earthquakes. Journal of Geophysical Research. 1980;**85**:3053-3057. DOI:10.1029/ JB085iB06p03053

[9] Virk HS, Manwinder Singh NP. Singh Surinder Singh (1986) calibration of radon detectors. International Journal of Radiation Applications and Instrumentation. Part D. Nuclear Tracks and Radiation Measurements. 1986;**12**(1-6):739-742

[10] Virk HS, Singh B. Radon anomalies in soil-gas and groundwater as earthquake precursor phenomena. Tectonophysics. 1993;**227**(1-4):215-224

[11] Virk HS, Singh B. Correlation of radon anomalies with earthquakes in the Kangra Valley. Journal of Nuclear Physics. 1992;**6**(2):293-300

[12] Virk HS. Uranium and radon surveys in western Himalayam. Current Science. 1997;**73**(25):536-538

[13] Segovia N, de la Cruz-Reyna S, Mena M, et al. Radon in soil anomaly observed at Los Azufres geothermal field, Michoacan: A possible precursor of the 1985 Mexico earthquake (Ms = 8.1). Natural Hazards. 1989;**1**:319-329. DOI: 10.1007/BF00134830

[14] King CY. Episodic radon changes in subsurface soil gas along active faults and possible relation to earthquakes. Journal of Geophysical Research. 1980;**85**:3065

[15] Akpan OU, Yakubu TA. A review of earthquake occurrences and observations

in earthquake science. Earthquake Science. 2010;**23**:289-294

[16] Yang TF, Walia B, Chyi LL, Fu CC, Chen CH, Liu TK, et al. Variations of soil radon and thoron concentrations in a fault zone and prospective earthquakes in SW Taiwan. Radiation Measurements. 1992;**40**(2-6):496-502

[17] Hartmann J, Levy JK. Hydrogeological and gas geochemical earthquake precursors – A review for application. Natural Hazards. 2005;**34**:279-304. DOI: 10.1007/s11069-004-2072-2

[18] Geologic and Mineral Resources Map of Nigeria. Nigeria: Geological Survey of Nigeria (GSNA); 2006

[19] Arabi AS, Zainab T, Funtua II, Mohammed AA, Ewa K, Abdulhamid MS, et al. Radon dynamics, activity concentration, and radiological risks associated with groundwater around tremor-prone Federal Capital Territory and environs, north-Central Nigeria. Arabian Journal of Geosciences. 2021;**14**:2479. DOI: 10.1007/s12517-021-08861-y

[20] RAD 7 Manual. Electric Radon Detector Operational Manual. Durridge Company Inc.; 2019

Chapter 7

Soil Contamination, Risk Assessment, and Remediation

Pooran Mal Meena, R.K. Aggarwal, Ramu Meena and Madhurjit Singh Rathore

Abstract

Soil amendment" refers to the alteration of soil properties driven by human activities. This includes actions such as spreading, compaction, erosion and fertility loss, which reshape the land and necessitate qualitative and quantitative risk assessment methods to address associated risks. Soil pollution, resulting from the excessive use of agrochemicals, waste materials, and toxic elements, disrupts soil organisms and fertility. Agrochemicals can contaminate soils through agricultural practices, impacting soil health. Irrigation water quality, indicated by salinity and soluble sodium percentage, also plays a crucial role in soil contamination. Additionally, emerging concerns arise from microplastics due to their persistence and potential ecological impacts. Remediation strategies for heavy metal-contaminated soils involve. Physical methods like soil replacement and thermal desorption address contamination by removing or treating soil on-site or off-site. Chemical fixation immobilizes contaminants using specific chemicals, while biological remediation, such as phytoremediation and bioremediation. Risk assessment, crucial for understanding the extent and severity of soil contamination, aids in effective remediation by considering factors such as contaminant concentrations, exposure pathways, and potential ecological and human health impacts. Overall, safeguarding soil health against pollution, erosion, and urbanization is essential for maintaining food and grazing resources.

Keywords: inorganic pollutants, trace elements, hazards, soil contamination, sewage and sludge, sustainability, soil microorganisms, heavy metals, pollutants, contamination, trace elements, hazards, sewage and sludge, fossil fuel, flora and fauna, sustainability, phytoremediation, solid waste, soil contamination, agrochemical, heavy metals

1. Introduction

Soil, a vital layer of organic and inorganic matter covering the earth's surface, sustains life by supporting plant growth and providing a habitat for various organisms [1]. The organic component, found in the uppermost layer of topsoil, consists of decaying plant and animal remains, while the inorganic part, formed through millennia of physical and chemical weathering, comprises crushed stone [1]. This dynamic environment is essential for food production, with 95% of human food originating from the soil [1]. However, soil pollution poses a significant threat, resulting from the introduction of exogenous chemicals or alterations to the natural soil environment, whether through natural processes or human activities [2].

Industrial expansion, mining activities, improper waste disposal, and agricultural practices contribute to soil contamination, with heavy metals and other pollutants accumulating over time [2, 3]. Common contaminants include lead, chromium, arsenic, zinc, cadmium, copper, mercury, and nickel, persisting in the soil for extended periods without degradation [4, 5]. The consequences of soil pollution are far-reaching, affecting human health, ecosystems, and food security [6, 7]. Direct exposure to contaminated soil, ingestion through the food chain, and contamination of groundwater resources pose significant risks [8]. Effective management of contaminated sites requires comprehensive risk assessment and the integration of scientific data to safeguard public and ecosystem health [9, 10].

Addressing soil contamination demands interdisciplinary expertise in environmental science, geology, hydrology, and chemistry [3]. While regulations in North America and Western Europe have made significant strides in identifying and mitigating soil pollution, challenges persist in regions with less stringent enforcement [11]. Enhanced awareness and proactive measures are essential to safeguard soil quality and mitigate the adverse effects of contamination worldwide.

2. Soil contamination

2.1 Processes of contamination

Trace elements are found in small levels in the environment and are ubiquitous under normal circumstances. The geological substrate is usually the main source of trace elements, with the oceans, soil, biota, and atmosphere coming in second and third. The relative distribution of the environmental divisions, for instance, is given for Like in Trace element concentrations in soil that are abnormally high and can be attributed to either natural or human sources. Volcanic activity, burning of wood or plants, and weathering (chemical and physical) processes are the natural sources of trace elements in soil. Anthropogenic sources of potentially hazardous trace elements in the environment are a result of urbanization and the Industrial Revolution [12]. These sources are connected to human activities such as the industrial production of steel, textiles, and chemicals; mining and smelter operations; burning of fossil fuels; incineration and disposal of trash; and agricultural practices (fertilizers and pesticides). The actions of humans have significantly accelerated the biogeochemical cycles of heavy metals. For a number of heavy metals, it has been calculated those human emissions to the atmosphere are one to three orders of magnitude greater than natural fluxes. Anthropogenic As, Sb, and Pb fluxes have been greater than natural fluxes for over 2000 years, according to research done on peat cores from a Swiss bog. The modification of the natural cycle of heavy metals has led to a situation in which the inputs of hazardous heavy metals in soils generally exceed the removal due to harvests of agricultural crops and the losses by leaching, volatilization, etc. [13].

2.2 Source of soil contaminations

1. Heavy Metals

2. Agrochemical

3. Irrigation water quality

4. Microplastics

2.3 Soil contamination of heavy metal

Soil contamination, characterized by the accumulation of toxic substances (pollutants and contaminants), poses significant risks to humans, soil organisms, animal health, and ecosystems. This contamination typically occurs in two primary forms: point pollution, originating from specific events at distinct locations like former industrial sites, and widespread pollution, where pollutants disperse over large areas, complicating detection and tracking efforts. Heavy metals infiltrate soil through natural phenomena and human activities, including industrial processes, mining, and agriculture. Understanding the sources, chemistry, and effects of heavy metal contamination is essential for devising effective conservation and remediation strategies.

2.3.1 Sources of heavy metals in contaminated soil

Agrochemicals (Pesticides, Fertilizers), Biosolids (Sewage Sludge), Wastewater. Pesticides, historically used extensively in agriculture and horticulture, contained significant concentrations of metals, contributing to soil contamination [11]. Fertilizers, essential for plant growth, often contain heavy metal impurities like cadmium and lead, which accumulate in soil over time [11, 14]. Similarly, biosolids, comprising municipal sewage and sludge, inadvertently introduce heavy metals into the soil through land application practices, raising environmental concerns [15, 16]. Additionally, long-term irrigation with municipal and industrial wastewater can result in heavy metal accumulation in soil [17].

2.3.2 Metal mining and milling processes and industrial wastes

Extensive mining and smelting activities lead to soil contamination with metals like lead and zinc, posing risks to human and ecological health [17]. Reclamation methods for these sites are often lengthy and expensive, with uncertain outcomes regarding soil productivity restoration. Soil heavy metal environmental risk to humans is associated with bioavailability, primarily through the ingestion of contaminated food or soil [17].

2.3.3 Soil concentration ranges for some heavy metals

The particular contamination of heavy metals that was found in the contaminated soil has a direct connection to the activity that took place there. Activities and disposal practices for contaminated wastes on the site will also affect the range of pollutant concentrations and their physical and chemical forms. Local transport pathways and the chemistry of the soil and groundwater are additional elements that could affect the type, concentration, and distribution of metal pollutants [5].

Soils may contain metals in the solid, gaseous, and liquid phases, which may complicate analysis and interpretation of reported results. For instance, total

Metals	Soil contamination range (mg kg $^{-1}$)	Regulatory limits (mg kg $^{-1}$)
Zn	150–5000	1500
Cr	0.05–3950	100
Pb	1.00–69,000	600
Cd	0.10–345	100
Hg	<0.01–1800	270

Table 1.
Ranges of soil concentrations and regulatory recommendations for specific heavy metals.

Metals	Target value (mg kg $^{-1}$)	Intervention value (mg kg $^{-1}$)
Cr	20	240
As	200	625
Ni	140.00	720.00
Cu	0.30	10.00
Zn	—	—
Cd	100.00	380.00
Pb	35.00	210.00
Hg	85	530

Table 2.
For several metals, the target and intervention values for a typical soil [18].

elemental analysis is the most widely used technique for figuring out how much heavy metal contamination is present in soil (USEPA Method 3050) (**Tables 1** and **2**).

2.3.4 Remediation of heavy metal-contaminated soils

Aims to create solutions protective of human health, animals, and the environment [19]. Regulatory compliance and risk assessments guide remediation strategies, emphasizing the reduction of metals' bioavailability to mitigate long-term risks [19]. Characterizing soil contamination determines remedial approaches, considering the physical form of heavy metal contaminants [20]. Technologies for remediation range from in-situ restrictive measures to gentle restoration methods, aiming to ensure safe soil use (**Table 3**).

2.4 Agrochemical

Agrochemicals or agricultural chemicals, short for agricultural chemicals, are chemicals used in agriculture. Agrochemicals refer to fungicides (pesticides include insecticides, herbicides, fungicides, and nematicides) and synthetic fertilizers. It also contains hormones and other growth factors [21].

Pesticides are products used to control pests [22]. These include (insecticides, herbicides, nematicides, fungicides, molluscicides, and plant growth regulators) [23]. Most of these are pesticides, accounting for approximately 50% of pesticides used worldwide [24]. Pesticides (insecticide, herbicide, fungicide, and nematicide) are

Category	Remediation technologies
Isolation	a. Capping b. Subsurface barriers
Immobilization	a. Solidification/stabilization b. Verification c. Chemical treatment
Toxicity and mobility reduction	a. Chemical treatment b. Permeable treatment c. Biological treatment bioaccumulation, phytoremediation (Phytoextraction, phytostabilization, and rhizofiltration), bioleaching, and biochemical processes
Extraction	a. Soil cleaning, electrokinetic treatment, pyrometallurgical extraction, and in-situ soil flushing

Table 3.
Technologies for remediation of heavy metals contaminated soils.

often used as plant protection products (also known as plant protection products) to protect plants from weeds, nematodes, fungi, or insects. Pesticides are generally defined as any chemical or biological material (such as bacteria, fungi, or viruses) that affects, inactivates, kills, or deters pests in any other way. Target pests include insects, parasites, plants, mollusks, birds, animals, fish, nematodes (roundworms), and diseases that damage crops, cause problems, or spread disease or disease. In addition to these benefits, pesticides also have disadvantages such as potential toxicity to humans and other species (**Figures 1** and 2).

2.4.1 Insecticides

Pesticides that kill insects are called insecticides [25]. These include larvicides and ovicides, which are used for insect larvae and eggs, respectively [26]. Agriculture makes use of insecticides. Almost all pesticides have the potential to drastically change ecosystems; many are harmful to the health of people and animals; some concentrate as they move up the food chain.

The development of the chemical industry at the same time made it easier to produce chlorinated hydrocarbons on a wide scale, including different forms of cyclodiene and hexachlorocyclohexane. Organophosphates are another big class of contact insecticides (**Tables 4** and 5).

2.4.1.1 Impact of insecticide on soil health/soil biology

Pesticide contamination [33] profoundly impacts soil ecosystems by disrupting microbial life crucial for soil health [34]. Insecticides, particularly carbamates, and organophosphates, have significant adverse effects on soil microbes [35]. Carbamate insecticides like carbofuran and carbaryl disrupt soil microbial environments and enzymatic activities [30, 36]. Similarly, organophosphates inhibit soil bacteria and fungi, while persistent chemicals like DDT, arsenic, and lindane further impair soil microbial biomass and enzymatic processes [31, 32, 37].

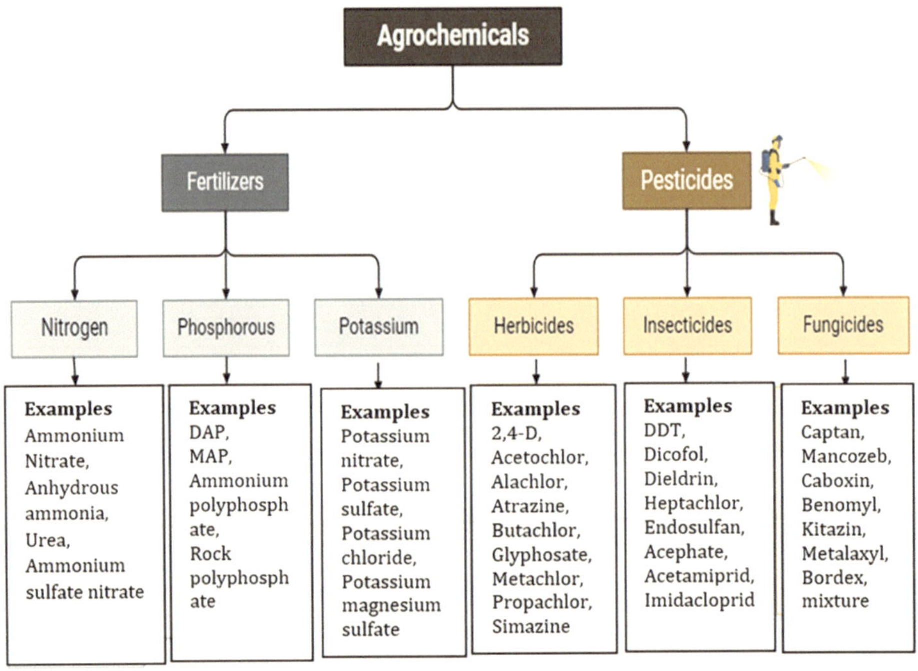

Figure 1.
An overview of agrochemical types: Classification and exemplary examples.

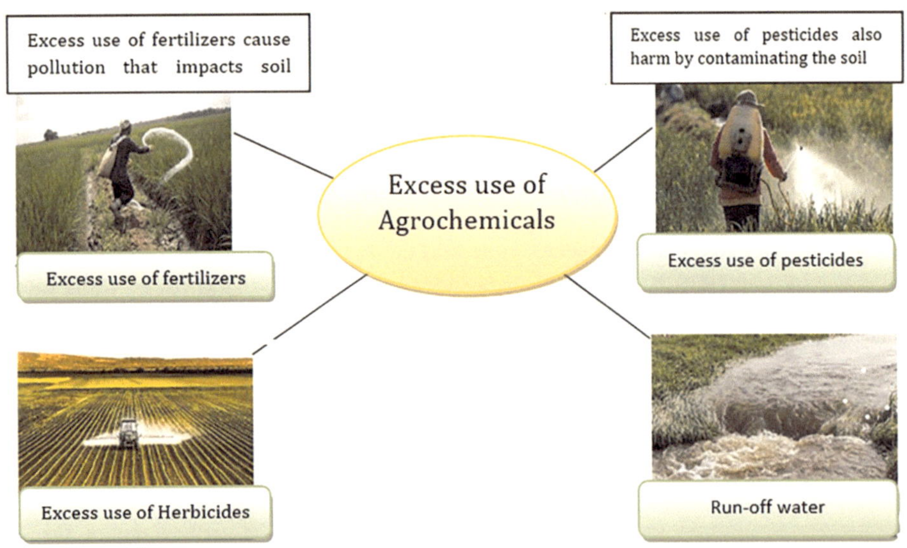

Figure 2.
Environmental and agricultural implications of excessive agrochemical use.

Unregulated insecticide application not only contaminates soil but also harms non-target organisms such as earthworms [38]. Studies reveal detrimental effects on earthworm biomass and cholinesterase activity due to pesticide exposure [39]. Malathion and chlorpyrifos exposure reduce earthworm body weight and reproductive viability, while cypermethrin diminishes cocoon production [40].

Sr. No.	Pesticides	Effect on soil microbes	References
1.	Imidacloprid and diazinon	The bacterium that produces urease is restrained (*Proteus vulgaris*)	[27]
2.	Chlorpyrifos. Quinalphos	Diminishes ammonification process	[28]
3.	Validamycin	Urease and phosphatase enzymes are negatively affected, but they improve subsequently	[29]
4.	Carbofuran	Hinders nitrogenase action of Anabaena doliolum was reduced by 38% Within 48 hours of treatment	[30]
5.	Organophosphate insecticide	Reduces the rate of N-mineralization by affecting the soil enzyme activity, the fungal species, and beneficial soil microbes	[31]
6.	DDT	As a result of their longer survival in soil. Microbial biomass production and enzyme reactions decreases	[32]

Table 4.
Impact of some insecticides on soil microbes.

Sr. No.	Name of Insecticide	Approximate Persistence time
1.	BHC (Benzene hexachloride)	11 years
2.	DDT	10 years
3.	Heptachlor	10 years
4.	Aldrin, Dieldrin	9 years
5.	Chlorodane	9 years
6.	Toxaphene	6 years

Table 5.
Persistence time for some selected insecticides.

Pesticides disrupt soil fertility and insecticide degradation processes, affecting local metabolism [41, 42].

2.4.2 Herbicides

Herbicides, essential for weed control, vary in selectivity [43]. Selective herbicides target specific weeds without harming desired crops, while non-selective types indiscriminately kill plants [44]. Integrated pest management strategies, combining pesticides with alternative pest control methods, offer sustainable solutions to mitigate pesticide-related soil damage and promote ecosystem resilience (**Tables 6** and 7).

2.4.2.1 Impact of herbicides on soil health (soil biology)

Herbicides reduce pathogen numbers within 7 to 30 days post-application, impacting microbial biodiversity [53]. Effects vary based on herbicide type and soil factors like texture and organic matter [54]. For instance, fomesafen herbicide

Sr. No.	Name of Herbicides	Approximate Persistence time
1.	Diuron	11 years
2.	Monuron	3 years
3.	Atrazine	18 years
4.	Simazine	17 years
5.	Chlordane	12 years
6.	2,3,6 - Trichlorobenzene	2-5 years
7.	2,4-D	2–8 years

Table 6.
Persistence time for some selected herbicides [45].

Sr. No.	Herbicides	Microorganisms, enzymes, and biochemical reactions	References
1.	2,4-D	Adversely affects the activities of Rhizobium spp. Reduces nitrogenase, phosphatase, and hydrogen photoproduction activities of purple non-sulfur bacteria	[46]
2.	2,4,5-T	Adversely impacts node-expression disrupting plant Rhizobium signaling. 2,4-D herbicide also reduces fixation by blue-green algae and nitrifying process impacting Nitrosomonas and *Nitrobacter* spp.	[47]
3.	Agroxone, and Atranex	Inhibits the most sensitive species of *Azotobacter vinelandii* and *Rhizobium phaseoli*	[46]
4.	Bromoxynil, Methomyl	Reduces CH_1 oxidation to CO_2	[48]
5.	Metsulfuron-methyl	Decreases N-mineralization	[49]
6.	Bentazone, Simazine,	Inhibits N-fixation and decreases the number of nodules and N content overall	[50]
7.	Isoproturon	Adverse consequences Actinomycetes and growth of fungi, urea-hydrolyzing bacteria, Nitrosomonas, Nitrobacter, and nitrate reductase activity	[51]
8.	Glyphosate	Reduces azotobacter development and activity	[36]
9.	Metribuzin	No impacts on AM fungus in maize and barley are seen at lower dosages	[52]

Table 7.
Herbicides and their reported effects on soil.

reduces soil microbial biomass carbon and mycorrhizal colonization after 12 days [55]. Triazine herbicides like atrazine alter soil microbial activity with prolonged use [56]. Glyphosate inhibits soil biota growth and phosphatase activity [36, 57].

2.4.3 Fungicides

Fungicides combat parasitic fungi, mitigating agricultural losses [58]. Contact fungicides protect sprayed areas, while systemic fungicides circulate through plant tissues [59]. Available in liquid or powder forms, fungicides contain active ingredients

Sr. No.	Fungicide	Effect on soil microbes	References
1.	Mancozeb	Effects of mycorrhizal associations and also nitrifying bacteria	[60]
2.	Benomyl	Hinders nodulation in legumes	[61]
3.	Carbendazim and Thiram	In soil, it affects bacteria which are associated with the N & C cycle	[62]
4.	Metalaxyl	Nitrifying, ammonifying bacteria's actions are disrupted	[63]
5.	Propiconazole	*Azospirillum brasilense*'s growth-promoting actions on the host plant can be inhibited	[64]
6.	Hexaconazole	Effects bacteria involved in N cycling	[65]
7.	Oxytetracycline	Acts as Bactericide	[66]

Table 8.
Effect of some fungicides on soil microbes.

ranging from 0.08–90% sulfur. They safeguard crops but can also impact soil health and biodiversity (**Table 8**).

Fungicides play a crucial role in agriculture by controlling fungal diseases, but they can have adverse effects on soil health [67]. Bavistin, a fungicide, exerts some inhibitory effects on soil microbial populations, albeit non-significantly [68]. While certain molecules in fungicides may affect Arbuscular Mycorrhizal Fungi (AMF), not all fungicides have uniform impacts on these fungi [69]. Benzoyl, found in some fungicides, can lead to long-term reductions in mycorrhizal associations [70], with many fungicides proving toxic to hyphal growth and root colonization by AMF [67]. Additionally, fungicides like emission and carbendazim can harm AMF populations [71].

Excessive fungicide use can lead to soil contamination and harm non-target organisms. Earthworms, vital for soil health, are negatively impacted by pesticides, with chlorpyrifos and azinphos methyl causing chronic and intermittent exposure effects [38, 39]. Other studies demonstrate reduced earthworm body weight and reproduction due to malathion exposure, highlighting the broader ecological repercussions of fungicide misuse [40]. Furthermore, fungicides disrupt soil fertility and the degradation processes crucial for ecosystem balance [41, 42].

2.4.4 Fertilizers

Fertilizers, both organic and inorganic, significantly influence soil properties [72]. Nitrogenous fertilizers, for instance, contribute to soil acidification, reducing pH levels and potentially limiting nutrient availability [73, 74]. Soil amendments enhance cation exchange capacity (CEC), which is crucial for nutrient retention and plant growth [75].

Atmospheric factors like acid rain also impact soil health [1, 76]. Acid rain, resulting from sulfur dioxide and nitrogen oxide emissions, can damage soil microorganisms, plants, and aquatic life [1]. It alters soil composition by depleting essential nutrients like calcium and magnesium, affecting plant growth and ecosystem dynamics [1, 77]. Additionally, acid rain corrodes human infrastructure, posing risks to public safety and architectural heritage [1, 78].

Balancing agricultural productivity with soil health is critical for sustainability [79]. Managing pesticide and fertilizer use, along with mitigating atmospheric pollutants, is essential for preserving soil fertility, ecosystem services, and human well-being [75]. Sustainable agricultural practices must prioritize soil conservation to ensure long-term food security and environmental resilience.

2.4.5 Acid deposition

Acid deposition, comprising wet and dry deposition [80], poses significant threats to soil health and ecosystem functioning. Wet deposition occurs when precipitation carries acids to the Earth's surface, while dry deposition involves the direct adherence of particles and gases to soil and plant surfaces [81]. Acid rain's adverse effects extend to forests, freshwater bodies, soil ecosystems, insects, and human health.

In soil, acid rain induces profound alterations in biology and chemistry. Low pH levels resulting from acid rain can kill sensitive bacteria and denature their enzymes, impairing soil microbial processes [82]. Additionally, hydronium ions in acid rain facilitate the leaching of toxins like lead and essential nutrients such as magnesium from soil [83]. This process disrupts soil nutrient balance and poses risks to plant health and growth. Acid rain also affects soil microbial communities, influencing nutrient cycling and soil fertility [84].

Soil acidification, driven by acid rain, exerts detrimental effects on plant physiology and growth [85]. Acidic water infiltrating plant tissues disrupts nutrient uptake and damages chloroplast organelles, hindering photosynthesis and nutrient production [78, 86]. Visible symptoms of soil acidification include yellowing leaves and reduced plant vigor [87]. Furthermore, acid rain compromises soil structure and nutrient availability, exacerbating soil degradation [88].

Mitigating soil acidification requires comprehensive strategies, including technical solutions and emission controls [89]. Power plants can implement gas-fired desulfurization to remove sulfur-containing gases, thereby reducing sulfur emissions responsible for acid rain formation [89]. Moreover, transitioning to alternative energy sources and enhancing vehicle emissions control can mitigate acid rain's environmental impacts [89].

2.5 Quality of irrigation water

Irrigation water quality: Another significant concern related to soil health is irrigation water quality [90]. Salinity, characterized by high salt concentrations in soil water, affects nutrient balance and plant growth, posing a global soil degradation threat [90, 91]. Saline and sodic soils cover vast areas globally, disrupting nutrient cycling and biological activity [92]. High sodium levels in soil solutions exacerbate clay dispersion and reduce nutrient uptake, impeding plant growth [93]. In arid and semi-arid climates, poor-quality irrigation water exacerbates soil salinization and sodification, diminishing soil fertility and agricultural productivity [94]. Addressing these challenges requires concerted efforts to improve water management practices and mitigate the impacts of salinity on soil ecosystems.

Considering the great importance of salinization and solidification of agricultural soil from an agro-environmental and economic point of view, it would be of interest to develop early diagnostic tools to control the safety of agricultural and environmental conditions [95]. One possibility is the use of infrared (IR) spectroscopy, a fast and powerful analytical technique often used to measure parameters such as soil carbon and nitrogen content [96]. Infrared spectroscopy is optional for soils by type or management [97]. Some studies on salinity and sodium estimation have used mapping techniques and satellite data [98], but few have used laboratory infrared spectroscopy to determine relative salinity and sodium in the ground.

Therefore, the purpose of this study is twofold: The first goal is to investigate the mid-term impacts on soil quality of irrigation water that carries a relative risk of salinity and sodicity. Second, to investigate if infrared spectroscopy can be used to measure the salinity and sodicity of various soil types.

2.5.1 US salinity laboratory staff classification

The term use salt-affected soil is being used more commonly to include saline, saline-sodic, and sodic soils (USSL Staff, 1954).

2.5.1.1 Saline soils

Saline soils are defined as soils that have a pH usually less than 8.5, electrical conductivity (EC) >4 dS m^{-1}, exchangeable sodium percentage (ESP) < 15, and sodium adsorption ratio (SAR) less than 13. A high electrical conductivity with a low ESP tends to flocculate soil particles into aggregates. When white salt crust is present for a portion of the area, it is typically easy to identify the soil types. The permeability is either higher than or comparable to that of typical, comparable soils [99, 100].

2.5.1.2 Saline-sodic soils

Saline-sodic soils contain sufficient soluble salts with electrical conductivity >4 dS m^{-1} to interfere with the growth of most crop plants and sufficient exchangeable sodium percentage (ESP) of more than 15 and sodium adsorption ratio (SAR) of more than 13 to affect the soil properties adversely, primarily by the degradation of soil structure. The pH may be more than 8.5 [100].

2.5.1.3 Sodic/alkali soils

Sodic soils exhibit an exchangeable sodium percentage (ESP) >15 and show an electrical conductivity (EC) <4 dS m^{-1}. The pH generally ranges between 8.5 and 10 and may be even as high as 11. The low ECe and high exchangeable sodium percentage (ESP) tend to deflocculate soil aggregates and, hence, lower their permeability to water [101].

2.5.1.4 Irrigation water quality standards

Salinity hazard or total salt concentration or EC: The concentration of soluble salt in irrigation water can be divided by electrical conductivity (EC) and expressed as dSm^{-1} [101] (USSL Staff 1954) (**Table 9**).

2.5.2 Salt index

The salt index is used to determine the stability of water in water containing more salt than sodium chloride. It is also used to estimate sodium levels. It is the relationship between Na^+, Ca_2^+, and $CaCO_3$ found in irrigation water. Total salt concentration expressed in ppm [101]:

$$\textbf{Salt Index} = (\text{Total Na} - 24.5) - [(\text{Total Ca} - \text{Ca in } CaCO_3) \text{ x } 4.85] \qquad (1)$$

The salinity test value for good water is negative; The saltwater test value is good because the water is not suitable for water use.

2.5.3 Sodium hazard (SAR)

Sodium hazards include high concentrations of Na + cations in water, which are undesirable because Na^+ adsorbs in the cation exchange zones of the soil, causing soil aggregates to break down and soil pores to close, preventing water from flowing. The tendency for sodium percentage to increase its proportion on the cation exchange site at the expense of other types of cations is estimated by the ratio of sodium content to the content of Ca + Mg in the irrigation water. This is called the Sodium Adsorption Ratio [102]:

$$\textbf{SAR}[Na]/[Ca][Mg]/21/2 \qquad (2)$$

Sodium Adsorption Ratio: It is the ratio of Na concentration to Ca + Mg concentration; all these concentrations are expressed in me/liter [USSL Staff 1954] (**Table 10**).

2.5.4 Bicarbonate hazards

The (Bicarbonate) HCO_3^- anion is an important hazardous anion in irrigation water. The residual sodium carbonate is used to evaluate the quantity of irrigation water (**Table 11**) [102]:

$$\textbf{RSC}\left(meL^{-1}\right) = \left(CO_3^{2-} + HCO_3^-\right) - \left(Ca^{2+} + Mg^{2+}\right) \qquad (3)$$

Salinity class	Salt concentration ($g\ L^{-1}$)	EC ($ds\ m^{-1}$)	Remarks
Low salinity (C_1)	<0.16	0.0–0.25	Suitable for safe usage irrigation purposes
moderate salinity (C_2)	0.16–0.50	0.25–0.75	Suitable for use with moderately leaching
High salinity (C_3)	0.50–1.50	0.75–2.25	can be used, with certain management techniques, for irrigation
Very high salinity (C_4)	1.50–3.00	2.25–5.00	Not suitable for use as irrigation

Table 9.
Salinity hazard or total soluble salt concentration or EC.

SAR classes	SAR values	Remarks
S_1 – Low Na^+	<10	Grown all kinds of crops and use water
S_2 – Medium Na^+	10–18	Water used for drainage in sandy soil
S_3 – High Na^+	18–26	Sensitive crops should not be taken
S_4 – Very high Na^+	>26	This water is not used for plants

Table 10.
Sodium concentration of irrigation water.

2.5.5 Chloride concentration

As electrical conductivity and sodium ions rise, so does the presence of chloride ions in irrigation water. Consequently, low-salinity waters contain these ions. In contrast to sodium ions, chloride ions have no effect on the physical characteristics of soil and are not adsorbed on it. As such, it has typically been left out of contemporary classification schemes. It does, however, play a role in several regional water classification systems (**Table 12**) [102]:

$$\textbf{Chloride concentration}\ (\text{me L}^{1}) = \text{Cl}^{-}/\text{CO}_3{}^{2-} + \text{HCO}_3{}^{-} + \text{SO}_4{}^{2-} + \text{CI}^{-} + \text{NO}_3{}^{-} \quad (4)$$

2.5.6 Soluble sodium percentage

Sodium is thought to be the most hazardous of the elements that are soluble in water. Too many sodium ions make water salty or alkaline depending on its combination with chloride/sulfate or carbonate/bicarbonate ions. In the past, Soluble Sodium Percentage (SSP), which may be computed as follows, was used to evaluate the sodium content in irrigation water [99]:

Water classes	RSC value	Remarks
Low RSC	<1.25	Suitable for safe usage
Medium RSC	1.25–2.50	Utilizable under specific management
High RSC	> 2.50	Not suitable for irrigation purposes

Table 11.
Bicarbonate concentration of irrigation water.

Chloride concentration (me L^{-1})	Water quality
4	Outstanding irrigation water
4–7	fairly good water
7–12	Slightly usable irrigation water
12–20	Unfit for use as irrigation water
>20	

Table 12.
Chloride concentration of irrigation water.

$$SSP\ (\%) = \frac{\text{Soluble Sodium (Na) Concentration}}{\text{Total Cation (Ca + Mg + Na) Concentration}} \times 100 \qquad (5)$$

Since a high value denotes soft water and a low value denotes hard water, it has proved helpful in characterizing water. A portion of the extra sodium-containing irrigation water (SSP = 66) is adsorbed on the soil. High salt irrigation water negatively affects plants and soils.

2.5.7 Magnesium hazard

It is believed that one of the important criteria in determining water quality is the magnesium content of all divalent cations, since the high adsorption of magnesium from the soil affects it. When the calcium–magnesium ratio drops below 50, negative effects may occur in the soil. Magnesium risk in irrigation water should have a Mg:Ca ratio [102]:

$$Mg - \text{Adsorption Ratio} = Mg^{2+}/Ca^{2+} + Mg^{2+} \qquad (6)$$

2.5.8 Nitrate concentration (me L^1)

Nitrate concentration (small L^1): Groundwater usually contains large amounts of nitrate. When this type of irrigation water continues into the soil, a large part of the soil's energy is affected, causing plants to deteriorate(**Table 13**) [102].

2.5.9 Reasons for salinity in soil

Salts in soils can have a variety of causes, the most frequent of which are enumerated below [103]:

- The natural saltiness of the soil (parent material)
- Salty and brackish water
- Limited outflow and an increasing water level
- Plant transpiration and surface evaporation
- Condensed vapors from sea sprays fall to the earth like rain
- Wind-borne salts that produce salty terrain

Water class	Nitrate Value (me L^1)	Remarks
Low Nitrate	5.0	Good water No problem
Medium Nitrate	5.0–30.0	Moderately good water
High Nitrate	> 30.0	Unsuitable for irrigation purposes

Table 13.
Nitrate concentration of irrigation water.

DOI: http://dx.doi.org/10.5772/intechopen.1005045

- Using fertilizers excessively
- Use of soil amendment materials
- Use of sewage sludge effluent
- Dumping of industrial brine onto the soil

2.5.10 Salt leaching

Salt leaching is a crucial technique for mitigating soil salinity and its associated problems [103]. Determining soil salinity levels requires the use of salt-free water, particularly devoid of sodium and calcium carbonate, to effectively reclaim saline soils. Adequate soil structure and drainage are prerequisites for efficient leaching. In cases of high soil salinity, multiple applications of freshwater may be necessary to leach salts below the root zone, coupled with proper drainage systems. However, in areas with shallow groundwater levels where leaching is impracticable, artificial drainage becomes indispensable, offering advantages in removing soil salinity even with low-quality water.

2.6 Microplastics

Microplastics, defined as plastic fragments less than 5 millimeters in length, pose significant environmental pollution [104]. They originate from various sources such as cosmetics, clothing, food packaging, and industrial processes. Two classifications of microplastics exist: primary and secondary. Primary microplastics are engineered particles, including microfibers, microbeads, and plastic pellets, while secondary microplastics result from the degradation of larger plastic items [104].

Primary microplastics find applications in various consumer products, including cosmetics, pharmaceuticals, and air jet technology, posing concerns due to their persistence and potential toxicity [104, 105]. Conversely, secondary microplastics, generated by the breakdown of larger plastics in oceans and soils, contribute significantly to environmental contamination [106]. Over time, physical, biological, and chemical processes degrade plastics into smaller particles, potentially impacting ecosystems and human health.

Nanoplastics, smaller than 1 μm, represent a growing concern due to their potential environmental threats. While their presence in the environment is increasingly recognized, uncertainties persist regarding their distribution and impact. Nanoplastics are speculated to accumulate in marine ecosystems, raising concerns about their effects on biodiversity and ecosystem functioning.

2.6.1 Sources and remediation of microplastic

Various sources contribute to microplastic pollution, including car tires, face masks, and sewage treatment plants [107, 108]. Tire wear is a significant contributor to soil microplastic contamination, necessitating measures to mitigate environmental impacts [109]. Similarly, disposable face masks, widely used during the COVID-19 pandemic, contribute to microplastic pollution upon degradation [110]. Sewage treatment plants also play a role in microplastic dissemination, with biosolids serving as a vector for microplastic transport into water bodies [106].

3. Risk assessment of soil contaminations

Oil spills represent another critical environmental concern, posing risks to marine ecosystems and human health [111]. Human activities, natural disasters, and equipment malfunctions contribute to oil spill occurrences, necessitating effective prevention and response measures [112]. Remediation efforts involve a combination of containment, cleanup, and risk assessment strategies to minimize environmental and socio-economic impacts [113].

In conclusion, salt leaching and microplastic pollution present significant environmental challenges, necessitating comprehensive strategies to mitigate their impacts on soil and aquatic ecosystems. Effective management practices, coupled with public awareness and regulatory interventions, are essential for addressing these complex environmental issues (**Figure 3**).

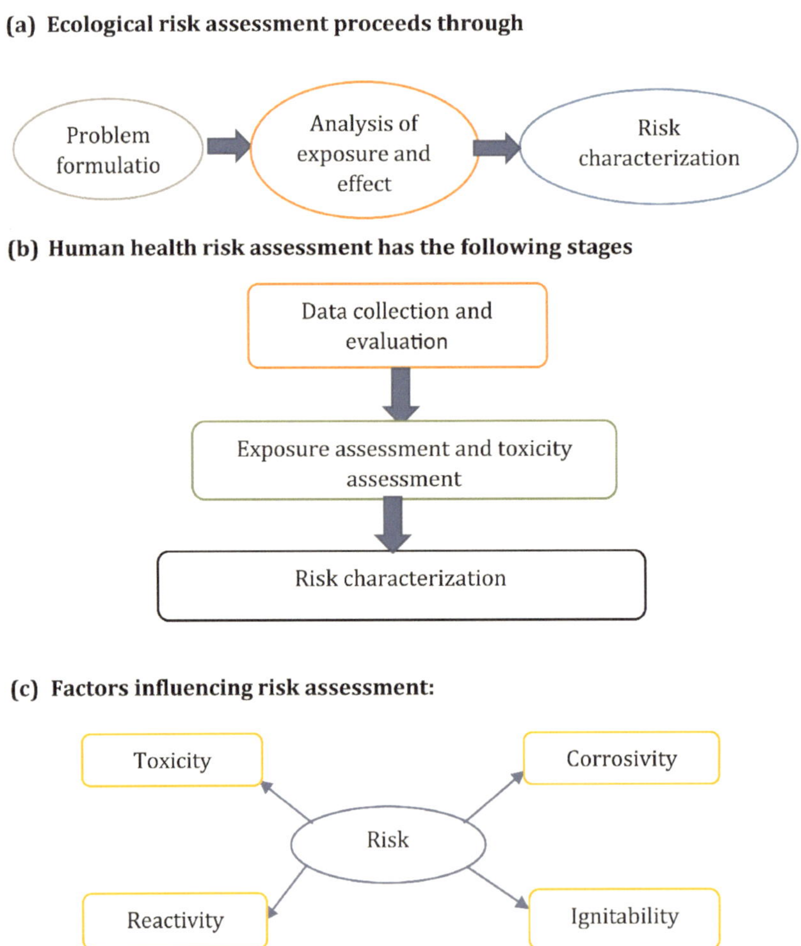

Figure 3.
The US EPA advocates performing HHRA and ERA through three stages. (a) Ecological risk assessment proceeds through. (b) Human health risk assessment has the following stages. (c) Factors influencing risk assessment.

4. Remediation of contaminated soils

Metals do not break down like carbon-based (organic) molecules. The only exceptions are mercury and selenium, which can be altered and replaced by bacteria. However, in general, it is very difficult to remove metals from the environment.

When large areas of soil are contaminated, traditional soil treatment methods can be used but are more costly. Treatment can be done on-site or off-site (off-site removal and treatment). Both are very expensive.

Some treatments include:

a. Highly temperature treatments

b. Solidifying agents (produce cement agent like material)

c. Washing out process (leaches out soil contaminants)

4.1 Remediation of heavy metal-contaminated soil

The impurities can be eliminated using different techniques, such as chemical, biological, and physical ones.

4.1.1 Physical methods

The physical remediation techniques are thermal desorption and soil replacement. Three different kinds of soil replacement exist:

- Replacing soil
- Importing new soil
- Spading soil

4.1.2 Chemical fixation

In this way, chemicals that reduce or prevent soil contamination are used.

4.1.3 Biological remediation

Phytoremediation, bioremediation, and combined remediation are examples of biological remediation.

a. **Bioremediation:** - (fungi, plants, green earth, bacteria) bioremediation of Contaminated Soil environmental remediation tackles pollution removal from soil, groundwater, and water bodies, safeguarding human health and ecosystems.

b. **Phytoremediation:** - Utilizing green plants to absorb, inhibit, and cleanse contaminants, phytoremediation offers a cost-effective solution (**Figure 4**).

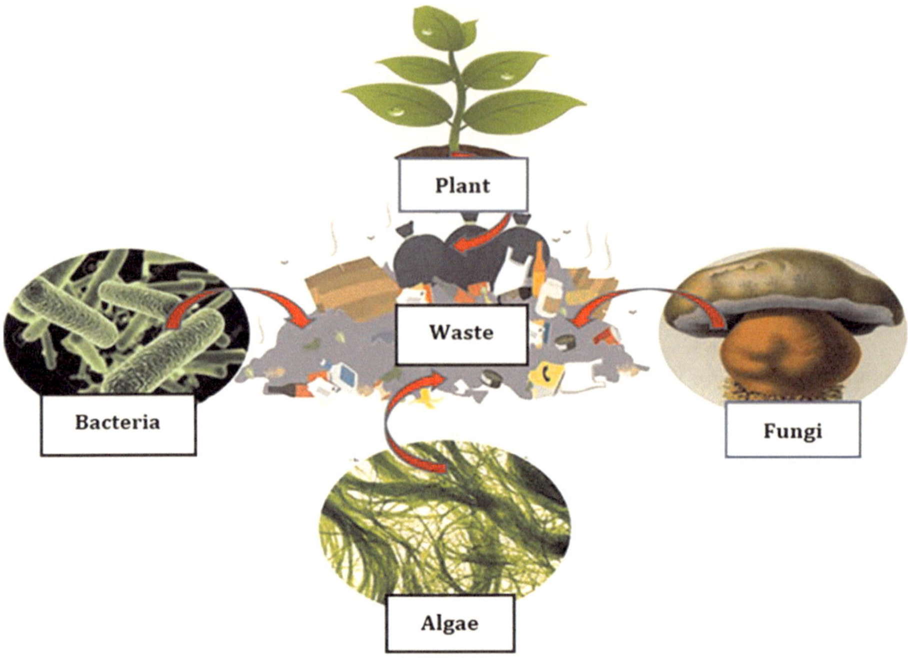

Figure 4.
Waste impact on plant health: Understanding the effects of soil contamination.

4.1.3.1 Phytoremediation techniques

a. Phytodegradation: Plants store and degrade contaminants.

b. Phyto-stimulation/Rhizo-degradation: Plant-microbe collaboration degrades contaminants.

c. Phyto-volatilization: Plants release contaminants as volatile compounds.

d. Phyto-extraction: Plants extract and degrade contaminants, especially heavy metals.

e. Phyto-stabilization: Plants reduce the mobility of heavy metals.

f. Rhizo-filtration: Roots filter contaminants from water sources.

5. Conclusion

In conclusion, soil amendment processes driven by human activities like agriculture and industry alter soil properties, posing risks such as pollution and fertility loss. Effective risk assessment methods are crucial for mitigating these threats and safeguarding soil health, essential for ecosystems, resource production, and human well-being. The intricate interplay of human activities and natural processes shapes soil properties, leading to contamination and significant threats to ecosystems and human well-being. Vital risk assessment methods aid in understanding and mitigating these risks, considering factors like contaminant concentrations, exposure pathways, and potential impacts. Remediation strategies, including physical, chemical, and biological methods, offer hope for restoring soil health and ecosystem integrity. Urgent action is needed to safeguard soil resources against pollution, erosion, and urbanization, ensuring sustainable food production and environmental health for future generations.

Acknowledgements

The authors express sincere gratitude to Dr. Yashwant Singh Parmar, University of Horticulture and Forestry, for their invaluable support and resources in the completion of this work.

Author details

Pooran Mal Meena[1*], R.K. Aggarwal[1], Ramu Meena[2] and Madhurjit Singh Rathore[1]

1 Dr. Yashwant Singh Parmar University of Horticulture and Forestry, Solan, Himachal Pradesh, India

2 Sri Karan Narendra Agriculture University, Jaipur, Rajasthan, India

*Address all correspondence to: pooranmeena301027@gmail.com

References

[1] Mandowara R. Acid rain-causes and effects. Journal of Science, Research and Teaching. 2023;**2**:9-32

[2] Khan S, Cao Q, Zheng YM, Huang YZ, Zhu YG. Health risks of heavy metals in contaminated soils and food crops irrigated with wastewater in Beijing, China. Environmental Pollution. 2008;**152**:686-692

[3] Zhang MK, Liu ZY, Wang H. Use of single extraction methods to predict bioavailability of heavy metals in polluted soils to rice. Communications in Soil Science and Plant Analysis. 2010;**41**: 820-831

[4] Evanko CR, Dzombak DA. Remediation of Metals-Contaminated Soils and Groundwater. Pittsburgh, PA, USA: Ground-water remediation technologies analysis center; 1997. pp. 5-13

[5] Kirpichtchikova TA, Manceau A, Spadini L, Panfili F, Marcus MA, Jacquet T. Speciation and solubility of heavy metals in contaminated soil using X-ray microfluorescence, EXAFS spectroscopy, chemical extraction, and thermodynamic modeling. Geochimica et Cosmochimica Acta. 2006;**70**: 2163-2190

[6] Ajmone-Marsan F, Biasioli M. Trace elements in soils of urban areas. Water, Air, & Soil Pollution. 2010;**213**:121-143

[7] McLaughlin MJ, Zarcinas BA, Stevens DP, Cook N. Soil testing for heavy metals. Communications in Soil Science and Plant Analysis. 2000;**31**: 1661-1700

[8] Ling W, Shen Q, Gao Y, Gu X, Yang Z. Use of bentonite to control the release of copper from contaminated soils. Soil Research. 2007; **45**:618-623

[9] Kabata-Pendias A, Pendias H. Trace Elements in Soil and Plants. Boca Raton, London, New-York, Washington, DC: CRC Press; 2001

[10] Zhao Q, Kaluarachchi JJ. Risk assessment at hazardous waste-contaminated sites with variability of population characteristics. Environment International. 2002;**28**:41-53

[11] Jones LHP, Jarvis SC. The fate of heavy metals. In: The Chemistry of Soil Processes. France: U.S. Department of Energy Office of Scientific and Technical Information; 1981. pp. 593-620

[12] Adriano DC, Chlopecka A. Soil Contamination and Remediation: Philosophy, Science and Technology. Wallingford, United Kingdom: CABI; 1997

[13] Jones KC. Contaminant trends in soils and crops. Environmental Pollution. 1991;**69**:311-325

[14] Lasat MM. Phytoextraction of metals from contaminated soil: A review of plant/soil/metal interaction and assessment of pertinent agronomic issues. Journal of Hazardous Substance Research. 1999;**2**:5

[15] Basta NT, Ryan JA, Chaney RL. Trace element chemistry in residual-treated soil: Key concepts and metal bioavailability. Journal of Environmental Quality. 2005;**34**:49-63

[16] United States. Environmental Protection Agency. Office of Wastewater Management, & United States. Environmental Protection

Agency. Office of Wastewater Management. Municipal Technology Branch. A Plain English Guide to the EPA Part 503 Biosolids Rule. US Environmental Protection Agency, Office of Wastewater Management; 1994

[17] Basta NT, Gradwohl R. Remediation of heavy metal-contaminated soil using rock phosphate. Better Crops. 1998;**82**: 29-31

[18] Egaspin DPR. Environmental Guidelines and Standards for the Petroleum Industry in Nigeria (EGASPIN). Lagos, Nigeria: Department of Petroleum Resources; 2002

[19] Martin TA, Ruby MV. Review of in situ remediation technologies for lead, zinc, and cadmium in soil. Remediation Journal: The Journal of Environmental Cleanup Costs, Technologies & Techniques. 2004;**14**:35-53

[20] Gupta SK, Herren T, Wenger K, Krebs R, Hari T. In situ gentle remediation measures for heavy metal-polluted soils. In: Phytoremediation of Contaminated Soil and Water. CRC Press; 2020. pp. 303-322

[21] "Agrochemicals and Security". University of Florida. Archived from the original on 2017-10-16. Retrieved 2008-12-14

[22] EPA, U. Basic Information about Pesticide Ingredients. US Environmental Protection Agency; 2018

[23] Randall C. Pest Management National Pesticide Applicator Certification Core Manual. 2014:24-26

[24] Atwood D, Paisley-Jones C. 2008–2012 Market Estimates. Pesticides Industry Sales and Usage. USEPA; 2017

[25] Dwivedi SA, Sonawane VK, Pandit TR. Review on the impact of insecticides utilization in crop ecosystem: Their prosperity and threats. In: Insecticides-Impact and Benefits of its Use for Humanity. London, UK: IntechOpen; 2022

[26] Van Emden HF, Peakall DB, International Centre of Insect Physiology and Ecology. Beyond Silent Spring: Integrated Pest Management and Chemical Safety. London: Chapman, Hall; 1996

[27] Ingram CW, Coyne MS, Williams DW. Effects of commercial diazinon and imidacloprid on microbial urease activity in soil and sod. Journal of Environmental Quality. 2005;**34**: 1573-1580

[28] Madhuri RJ, Rangaswamy V. Influence of selected insecticides on phosphatase activity in groundnut (*Arachis hypogaea* L.) soils. Journal of Environmental Biology. 2002;**23**:393-397

[29] Qian H, Hu B, Wang Z, Xu X, Hong T. Effects of validamycin on some enzymatic activities in soil. Environmental Monitoring and Assessment. 2007;**125**:1-8

[30] Kalam A, Mukherjee AK. Influence of hexaconazole, carbofuran and ethion on soil microflora and dehydrogenase activities in soil and intact cell. Indian Journal of Experimental Biology. 2001; **39**(1):90-94

[31] Pandey S, Singh DK. Total bacterial and fungal population after chlorpyrifos and quinalphos treatments in groundnut (*Arachis hypogaea* L.) soil. Chemosphere. 2004;**55**:197-205

[32] Singh J, Singh DK. Dehydrogenase and phosphomonoesterase activities in groundnut (*Arachis hypogaea* L.) field

after diazinon, imidacloprid and lindane treatments. Chemosphere. 2005;**60**: 32-42

[33] Zhu G, Wu H, Guo J, Kimaro FM. Microbial degradation of fipronil in clay loam soil. Water, Air, and Soil Pollution. 2004;**153**:35-44

[34] Shang Y, Hasan MK, Ahammed GJ, Li M, Yin H, Zhou J. Applications of nanotechnology in plant growth and crop protection: A review. Molecules. 2019;**24**:2558

[35] Aktar MW, Sengupta D, Chowdhury A. Impact of pesticides use in agriculture: Their benefits and hazards. Interdisciplinary Toxicology. 2009;**2**:1

[36] Sannino F, Gianfreda L. Pesticide influence on soil enzymatic activities. Chemosphere. 2001;**45**:417-425

[37] Van Zwieten L, Ayres MR, Morris SG. Influence of arsenic co-contamination on DDT breakdown and microbial activity. Environmental Pollution. 2003;**124**:331-339

[38] Azam F, Farooq S, Lodhi A. Microbial biomass in agricultural soils-determination, synthesis, dynamics and role in plant nutrition. Pakistan Journal of Biological Sciences (Pakistan). 2003;**6**: 629-639

[39] Reinecke SA, Reinecke AJ. The impact of organophosphate pesticides in orchards on earthworms in the Western cape, South Africa. Ecotoxicology and Environmental Safety. 2007;**66**: 244-251

[40] Yasmin S, D'Souza D. Effects of pesticides on the growth and reproduction of earthworm: A review. Applied and Environmental Soil Science. 2010;**2010**:678360

[41] Topp E, Vallaeys T, Soulas G. Pesticides: Microbial degradation and effects on microorganisms. Modern Soil Microbiology. 1997:547-575

[42] Andrea MM, Peres TB, Luchini LC, Marcondes MA, Pettinelli A Jr, Nakagawa LE. Impact of long term applications of cotton pesticides on soil biological properties, dissipation of 14 C-methyl parathion and persistence of multi-pesticide residues No. IAEA-TECDOC–1248. In: Proceedings of a Final Research Coordination Meeting. Vienna, Austria: International Atomic Energy Agency (IAEA); 2001

[43] Sales PI. Usage: 2006 and 2007 Market Estimates. US Environmental Protection Agency; 2011

[44] Ura Y, Sakata G. Ullmann's Encyclopedia of Industrial Chemistry. Wiley; 2001

[45] Meena RS, Kumar S, Datta R, Lal R, Vijayakumar V, Brtnicky M, et al. Impact of agrochemicals on soil microbiota and management: A review. Land. 2020;**9**:34

[46] Fabra A, Duffard R, De Duffard AE. Toxicity of 2, 4-dichlorophenoxyacetic acid to rhizobium sp in pure culture. Bulletin of Environmental Contamination and Toxicology. 1997;**59**: 645-652

[47] Fox JE, Starcevic M, Kow KY, Burow ME, McLachlan JA. Nitrogen fixation: Endocrine disrupters and flavonoid signalling. Nature. 2001;**413**: 128-130

[48] Syamsul Arif MA, Houwen F, Verstraete W. Agricultural factors affecting methane oxidation in arable soil. Biology and Fertility of Soils. 1996; **21**:95-102

[49] Subhani A, El-ghamry AM, Changyong H, Jianming X. Effects of pesticides (herbicides) on soil microbial biomass-a review. Pakistan Journal of Biological Sciences. 2000;**3**:705-709

[50] Singh G, Wright D. In vitro studies on the effects of herbicides on the growth of rhizobia. Letters in Applied Microbiology. 2002;**35**:12-16

[51] Nowak J, Kaklewski K, Klódka D. Influence of various concentrations of selenic acid on the activity of soil enzymes. Science of the Total Environment. 2002;**291**:105-110

[52] Makarian H, Poozesh V, Asghari HR, Nazari M. Interaction effects of arbuscular mycorrhiza fungi and soil applied herbicides on plant growth. Communications in Soil Science and Plant Analysis. 2016;**47**:619-629

[53] Milošević NA, Govedarica MM. Effect of herbicides on microbiological properties of soil. Zbornik Matice srpske za prirodne nauke. 2002;**102**:5-21

[54] Kremer RJ, Means NE. Glyphosate and glyphosate-resistant crop interactions with rhizosphere microorganisms. European Journal of Agronomy. 2009;**31**:153-161

[55] Santos JB, Jakelaitis A, Silva AA, Costa MD, Manabe A, Silva MCS. Action of two herbicides on the microbial activity of soil cultivated with common bean (Phaseolus vulgaris) in conventional-till and no-till systems. Weed Research. 2006;**46**:284-289

[56] Seghers D, Verthé K, Reheul D, Bulcke R, Siciliano SD, Verstraete W, et al. E_ect of long-term herbicide applications on the bacterial community structure and function in an agricultural soil. FEMS Microbiology Ecology. 2003; **46**:139-146

[57] Santos A, Flores M. Effects of glyphosate on nitrogen fixation of free-living heterotrophic bacteria. Letters in Applied Microbiology. 1995;**20**:349-352

[58] Dreikorn BA, Owen WJ. Fungicides, Agricultural. In: Kirk-Othmer Encyclopedia of Chemical Technology. Wiley; 2000

[59] Latijnhouwers M, de Wit PJ, Govers F. Oomycetes and fungi: Similar weaponry to attack plants. Trends in Microbiology. 2003;**11**:462-469

[60] Černohlávková J, Jarkovský J, Hofman J. Effects of fungicides mancozeb and dinocap on carbon and nitrogen mineralization in soils. Ecotoxicology and Environmental Safety. 2009;**72**:80-85

[61] Chen SK, Edwards CA, Subler S. Effects of the fungicides benomyl, captan and chlorothalonil on soil microbial activity and nitrogen dynamics in laboratory incubations. Soil Biology and Biochemistry. 2001;**33**:1971-1980

[62] Niewiadomska A. Effect of Carbendazim, Imazetapir and Thiram on nitrogenase activity, the number of microorganisms in soil and yield of red clover Trifolium pratense L. Polish Journal of Environmental Studies. 2004; **13**:403-410

[63] Monkiedje A, Spiteller M. Degradation of metalaxyl and mefenoxam and effects on the microbiological properties of tropical and temperate soils. International Journal of Environmental Research and Public Health. 2005;**2**:272-285

[64] Pereyra MA, Ballesteros FM, Creus CM, Sueldo RJ, Barassi CA. Seedlings growth promotion by Azospirillum brasilense under normal

and drought conditions remains unaltered in tebuconazole-treated wheat seeds. European Journal of Soil Biology. 2009;**45**:20-27

[65] Madhuri RJ, Rangaswamy V. Influence of selected fungicides on microbial population in groundnut (*Arachis hypogeae* L.) soils. Pollution Research. 2003;**22**:205-212

[66] Qingxiang YANG, Zhang J, Kongfang ZHU, Zhang H. Influence of oxytetracycline on the structure and activity of microbial community in wheat rhizosphere soil. Journal of Environmental Sciences. 2009;**21**:954-959

[67] Cycoń M, Piotrowska-Seget Z, Kaczyńska A, Kozdrój J. Microbiological characteristics of a sandy loam soil exposed to tebuconazole and λ-cyhalothrin under laboratory conditions. Ecotoxicology. 2006;**15**:639-646

[68] Anil Gupta AG, Ashok Aggarwal AA, Chhavi Mangla CM, Aditya Kumar AK, Anju Tanwar AT. Effect of herbicides fenoxaprop-P-ethyl and 2, 4-D ethyl-ester on soil mycoflora including VAM fungi in wheat crop. Indian Journal of Weed Science. 2012;**43**(1 & 2):32-40

[69] Datta R, Anand S, Moulick A, Baraniya D, Imran Pathan S, Rejsek K, et al. How enzymes are adsorbed on soil solid phase and factors limiting its activity: A review. International Agrophysics. 2017;**31**:287-302

[70] Smith MD, Hartnett DC, Rice CW. Effects of long-term fungicide applications on microbial properties in tallgrass prairie soil. Soil Biology and Biochemistry. 2000; **32**:935-946

[71] Murillo-Williams A, Pedersen P. Arbuscular Mycorrhizal colonization response to three seed-applied fungicides. Agronomy Journal. 2008;**100**:795

[72] Scherer HW et al. Fertilizers, 1. General. In: Ullmann's Encyclopedia of Industrial Chemistry. Wiley; 2000

[73] Schindler DW, Hecky RE. Eutrophication: more nitrogen data needed. Science. 2009;**324**:721-722

[74] Penn CJ, Bryant RB. Phosphorus solubility in response to acidification of dairy manure amended soils. Soil Science Society of America Journal. 2008;**72**: 238-243

[75] Ozlu E, Kumar S. Response of soil organic carbon, pH, electrical conductivity, and water stable aggregates to long-term annual manure and inorganic fertilizer. Soil Science Society of America Journal. 2018;**82**: 1243-1251

[76] Blair T. Diversifying Clean Water an Examination of Drinking Water Quality and Social Disparities in Michigan Doctoral Dissertation. USA: Tennessee State University; 2021

[77] Magaino S. Corrosion rate of copper rotating-disk-electrode in simulated acid rain. Electrochimica Acta. 1997;**42**: 377-382

[78] Markewit D, Richter DD, Allen HL, Urrego JB. Three decades of observed soil acidification in the Calhoun experimental forest: Has acid rain made a difference. Soil Science Society of America Journal. 1998;**62**:1428-1439

[79] Pahalvi HN, Rafiya L, Rashid S, Nisar B, Kamili AN. Chemical fertilizers and their impact on soil health. In: Microbiota and Biofertilizers. Ecofriendly Tools for Reclamation of Degraded Soil Environments. Vol. 2. Switzerland: Springer Nature; 2021. pp. 1-20

[80] Lidsky TI. Is the aluminum hypothesis dead. Journal of Occupational and Environmental Medicine. 2014;**56**:73

[81] Kesler SE, Simon AF. Mineral Resources, Economics and the Environment. Cambridge University Press; 2015

[82] Rodhe H, Dentener F, Schulz M. The global distribution of acidifying wet deposition. Environmental Science & Technology. 2002;**36**:4382-4388

[83] Likens GE, Driscoll CT, Buso DC. Long-term effects of acid rain: Response and recovery of a forest ecosystem. Science. 1996;**272**:244-246

[84] Zhang Y, Li J, Tan J, Li W, Singh BP, Yang X, et al. An overview of the direct and indirect effects of acid rain on plants: Relationships among acid rain, soil, microorganisms, and plants. Science of the Total Environment. 2023;**873**:162388

[85] Larssen T, Carmichael GR. Acid rain and acidification in China: The importance of base cation deposition. Environmental Pollution. 2000;**110**:89-102

[86] Evans LS, Gmur NF, Costa FD. Leaf surface and histological perturbations of leaves of *Phaseolus vulgaris* and *Helianthus annuus* after exposure to simulated acid rain. American Journal of Botany. 1977;**64**:903-913

[87] Du YJ, Wei ML, Reddy KR, Liu ZP, Jin F. Effect of acid rain pH on leaching behavior of cement stabilized lead-contaminated soil. Journal of Hazardous Materials. 2014;**271**:131-140

[88] Sun J, Hu H, Li Y, Wang L, Zhou Q, Huang X. Effects and mechanism of acid rain on plant chloroplast ATP synthase. Environmental Science and Pollution Research. 2016;**23**:18296-18306

[89] Stoyanova D, Velikova V. Effects of simulated acid rain on chloroplast ultrastructure of primary leaves of *Phaseolus vulgaris*. Biologia Plantarum. 1997;**40**:589-595

[90] Shahid SA, Mahmoudi H. National strategy to improve plant and animal production in the United Arab Emirates. Soil and Water Resources Annexes. 2014: 113-131

[91] Zaman M, Shahid SA, Heng L, Zaman M, Shahid SA, Heng L. Irrigation water quality. In: Guideline for Salinity Assessment, Mitigation and Adaptation Using Nuclear and Related Techniques. Springer; 2018. pp. 113-131

[92] Daliakopoulos IN, Tsanis IK, Koutroulis A, Kourgialas NN, Varouchakis AE, Karatzas GP, et al. The threat of soil salinity: A European scale review. Science of the Total Environment. 2016;**573**:727-739

[93] Rengasamy P. World salinization with emphasis on Australia. Journal of Experimental Botany. 2006;**57**:1017-1023

[94] Grattan SR, Grieve CM. Salinity–mineral nutrient relations in horticultural crops. Scientia Horticulturae. 1998;**78**:127-157

[95] Herrero J, Snyder RL. Aridity and irrigation in Aragon, Spain. Journal of Arid Environments. 1997;**35**:535-547

[96] Soriano-Disla JM, Janik LJ, Viscarra Rossel RA, Macdonald LM, McLaughlin MJ. The performance of visible, near-, and mid-infrared reflectance spectroscopy for prediction of soil physical, chemical, and biological properties. Applied Spectroscopy Reviews. 2014;**49**:139-186

[97] Aranda V, Domínguez-Vidal A, Comino F, Calero J, Ayora-Cañada MJ.

Agro-environmental characterization of semi-arid Mediterranean soils using NIR reflection and mid-IR-attenuated total reflection spectroscopies. Vibrational Spectroscopy. 2014;**74**:88-97

[98] Castrignanò A, Buttafuoco G, Puddu R. Multi-scale assessment of the risk of soil salinization in an area of South-Eastern Sardinia (Italy). Precision Agriculture. 2008;**9**:17-31

[99] Gupta RK, Abrol IP. Salt-affected soils: Their reclamation and management for crop production. Advances in Soil Science: Soil Degradation. 1990;**11**:223-288

[100] Regional Salinity Laboratory (US). Diagnosis and Improvement of Saline and Alkali Soils No. 60. US Department of Agriculture; 1954

[101] Szabolcs I. In: Nijhoff M, editor. Salt Affected Soils in Europe. Research Institute for Soil Science and Agricultural Chemistry of the Hungarian Academy of Sciences; 1974

[102] Eaton FM. Significance of carbonates in irrigation waters. Soil Science. 1950;**69**:123-134

[103] Abou-Baker NHA, El-Dardiry EA. Integrated Management of Salt Affected Soils in Agriculture: Incorporation of Soil Salinity Control Methods. Academic Press; 2015

[104] Ghosh S, Sinha JK, Ghosh S, Vashisth K, Han S, Bhaskar R. Microplastics as an emerging threat to the global environment and human health. Sustainability. 2023;**15**(14):10821

[105] Karbalaei S, Hanachi P, Walker TR, Cole M. Occurrence, sources, human health impacts and mitigation of microplastic pollution. Environmental Science and Pollution Research. 2018;**25**: 36046-36063

[106] Cole M, Lindeque P, Halsband C, Galloway TS. Microplastics as contaminants in the marine environment: A review. Marine Pollution Bulletin. 2011;**62**:2588-2597

[107] Lassen C, Hansen SF, Magnusson K, Hartmann NB, Jensen PR, Nielsen TG, et al. Microplastics: Occurrence, Effects and Sources of Releases to the Environment in Denmark. Danish Environmental Protection Agency; 2015

[108] Saliu F, Veronelli M, Raguso C, Barana D, Galli P, Lasagni M. The release process of microfibers: From surgical face masks into the marine environment. Environmental Advances. 2021;**4**:100042

[109] Burghardt TE, Pashkevich A, Babić D, Mosböck H, Babić D, Żakowska L. Microplastics and road markings: The role of glass beads and loss estimation. Transportation Research Part D: Transport and Environment. 2022;**102**:103123

[110] Fadare OO, Okoffo ED. Covid-19 face masks: A potential source of microplastic fibers in the environment. The Science of the Total Environment. 2020;**737**:140279

[111] Diwakar S, Rajkumar K. Preparation of super hydrophobic loofah sponge for fast and efficient separation of oil from seawater. Materials Today: Proceedings. 2018;**5**:14367-14374

[112] Michel J, Fingas M. Oil spills causes, consequences, prevention, and countermeasures. In: Fossil Fuels: Current Status and Future Directions. World Scientific Publishing Company Pte. Limited; 2016. pp. 159-201

[113] Nandagopal KR, Kumar MP. Oil Contaminated sea water recovery system using composite textile absorption media. International Journal of Engineering Research. 2016;**5**:311-314

Chapter 8

The Implications for Risk Management in the Era of Technological Advancements

Monument Thulani Bongani Makhanya

Abstract

Amidst a period characterised by swift technological progress, risk management encounters unparalleled obstacles and prospects. The many facets of this paradigm change are examined in this paper. Conventional risk assessment techniques need to change as businesses are revolutionised by technologies like blockchain, IoT, and artificial intelligence. Even though these advances increase production and efficiency, they also bring new vulnerabilities, which means risk profiles need to be reevaluated. Furthermore, cascading risks are made more likely by the growing interconnection of global systems. Cybersecurity becomes critical, necessitating advanced precautions to protect private data. Moreover, new instruments for risk prediction and mitigation are made possible by the combination of machine learning and predictive analytics. The ethical implications of automated decision-making, on the other hand, necessitate careful examination. Organisations must promote adaptability in this volatile terrain by fostering a culture of constant learning and innovation. Navigating these difficulties effectively will define an enterprise's resilience and durability in a digitally driven future. This chapter explores the implications of risk management in the era of technological advancements and how those risks could be mitigated. The methodology employed in this chapter was secondary sources, and the gathered data was evaluated using text content to generate key insights.

Keywords: risk management, technology, risk assessments and mitigation, culture of learning and innovation, ethical implications

1. Introduction

The swift progress of technology has significantly affected enterprises, organisations, and the community at large. Advancements in many domains including artificial intelligence, blockchain, the Internet of Things, and robots have revolutionised the ways in which businesses function, organisations are structured, and individuals engage with one another [1]. Mohd et al. [2] suggest that increased productivity and efficiency for businesses and organisations is one of the main effects of the rapid improvements in technology. Automated and robotic systems have made it possible to complete tedious and repetitive activities faster and with higher accuracy. Human error is decreased as a result, and human resources are freed up to work on more intricate and important projects. Robotics adoption, for instance, has accelerated production lines, decreased labour costs, and enhanced product quality in the manufacturing sector. Technology has also completely changed how people communicate and work together in organisations. With the advent of numerous technological tools, including instant messaging apps, video conferencing software, and project management systems, workers may collaborate easily from anywhere in the world. This has made it easier for flexible work schedules and remote work to proliferate, giving businesses access to a worldwide talent pool, cutting expenses, and improving employee work-life balance [3].

According to Gupta et al. [4], the dynamic between firms and consumers has also been changed by technological improvements. The retail sector has changed as a result of the internet and e-commerce, which have allowed companies to expand their customer base and run around the clock. Nowadays, shoppers may quickly evaluate products, compare costs, and purchase them all from the convenience of their homes. Businesses have been compelled by this change to adjust and adopt digital strategies in order to guarantee that their online presence is optimised and that customers can easily access their goods and services. Rapid technical improvements have also had a major impact on information availability and accessibility. The internet has expanded into a massive knowledge base that makes it possible for people to learn new skills and obtain information at any time. This has made education more accessible to all people and given them the chance to advance both personally and professionally, regardless of where they live. This enables companies to access highly trained and informed staff, leading to more creative and competitive firms [5]. The quick speed at which technology is developing demands constant learning and adaptability. In order to adopt new technologies and incorporate them into their operational procedures, organisations need to be proactive and flexible. Those that fall behind run the risk of becoming obsolete or being surpassed by more creative rivals [6].

To that effect, companies and organisations must manage a wide range of novel and complicated risks, which, if ignored, might have dire repercussions. Emerging risks have the potential to impair operations, harm reputations, and even result in financial loss. These risks range from cybersecurity attacks to legislative changes. For this reason, proactive risk identification and management is essential to an organisation's long-term viability and performance [7]. The intrinsic vulnerability of technological systems is one of the main justifications for the significance of recognising and controlling new threats. Businesses are more vulnerable to cybersecurity assaults as a result of their growing reliance on technology. Cybercriminals and hackers are always coming up with new ways to take advantage of holes in digital infrastructure, which can lead to financial loss, reputational harm, and data breaches. Organisations should keep one step ahead of any threats and put strong security measures in place to safeguard their systems and sensitive data by recognising new risks in cybersecurity [8]. In the current tech world, where new technologies are always being invented and implemented, innovation and disruption are frequent. These developments, nonetheless, frequently carry risks and uncertainties that provide problems for businesses. When automation and artificial intelligence are combined, for instance, productivity may rise, but there may also be job losses and moral dilemmas. Organisations can optimise their utilisation of novel technology while simultaneously minimising any possible adverse effects by proactively recognising and addressing these developing risks [9].

Regulatory compliance makes it clear how important it is to manage new risks. Laws and regulations are always changing to keep up with the way that technology is reshaping economies and industries. To maintain compliance with constantly evolving standards, organisations need to be on the lookout for emerging regulatory risks. If you do not, you risk facing legal repercussions, harm to your reputation, and erosion of public confidence. Organisations may stay out of legal hot water and have good relations with regulators and stakeholders by proactively detecting and mitigating potential regulatory risks [10]. Moreover, according to Farida and Setiawan [11], in the technologically advanced world, controlling new risks is crucial to keeping a competitive advantage. The business world of today is extremely dynamic, and companies that remain ahead of new threats are better able to change and grow. Organisations may foresee possible disruptions, exploit opportunities, and maintain an advantage over competitors by being proactive in recognising and managing developing risks. It eventually positions individuals for long-term success by empowering them to take calculated risks and make smart decisions that are in line with their overarching goals. Managing emerging risks also helps companies develop an agile and resilient culture. Organisations can create strong frameworks for risk management that enable them to effectively handle possible crises by proactively monitoring and resolving developing threats. By doing this, any losses are lessened and the effects of any disruptions on stakeholders, customers, and operations are also reduced. Through fostering a culture of risk consciousness and readiness, establishments can acquire the adaptability required to manoeuvre through ambiguities and emerge more robust from hardship [12]. It is against this background that this chapter seeks to explore the implications for risk management in the era of technological advancements and how those risks could be mitigated.

2. Methodology

This chapter utilised secondary sources as methodology. Secondary sources offer researchers knowledge and data that has already been gathered and examined by others, making them useful as a research methodology. This is achieved by examining academic books, papers, essays, and other published resources that address and analyse the subject of interest [13]. According to Goundar [14], scholars scrutinise secondary sources to investigate extant knowledge and theoretical frameworks associated with their research subject. As a result, they are better able to formulate research questions and hypotheses and find gaps in the literature.

Researchers can obtain data that has already been gathered by others by using secondary sources. The researcher's own data can be compared with this data or used for additional analysis. To make inferences or bolster the results of the current investigation, statistical data from government agencies or earlier research, for instance, can be examined [15]. Secondary sources can be used by researchers to contrast and compare their results with those of earlier investigations. This adds to the body of knowledge in the field and supports the validity and generalizability of the research findings [16]. Historical researchers frequently employ secondary sources to comprehend and analyse historical events. Scholars examine historical viewpoints, trends, and patterns through the use of pre-published books, papers, and documents [17]. According to Ahn and Kang [18] in meta-analyses and systematic reviews, secondary sources are employed to combine the results of several primary investigations. Researchers are able to uncover common trends or patterns and develop more thorough conclusions by merging and analysing the data from multiple investigations.

To that effect, Secondary sources were chosen to acquire information on the implications for risk management in the age of technology breakthroughs mostly due to their accessibility and credibility. Sources, such as research articles, industry reports, books, databases, and online platforms, were used. In addition, the researcher deemed that these secondary sources were generated by specialists and have undergone thorough peer review, assuring their dependability and authenticity. They gave the researcher thorough coverage of the topic by synthesising information from diverse primary sources, which allowed the researcher to acquire a greater grasp of the implications of technological improvements on risk management practises.

3. New risks associated with technology development

According to Păvăloaia and Necula [19], technological advances have transformed how we live, work, and interact. These advancements, ranging from artificial intelligence and robots to nanotechnology and gene editing, have created enormous prospects and benefits. They do, however, bring with them new risks that pose enormous difficulties to individuals, organisations, and society as a whole. The threat presented by cyberattacks and data breaches is one of the most important rising concerns. Malicious actors may be able to take advantage of weaknesses in our systems as our dependence on technology increases. Cyberattacks have the potential to endanger national security, cause financial losses, steal confidential data, and damage vital infrastructure. Cybercriminals' techniques also advance with technology, so it is critical for organisations to invest in strong cybersecurity measures to safeguard their assets [20].

The ethical consequences of developing technology are another growing risk. As technological breakthroughs such as artificial intelligence and robotics continue to advance, the ethical issues surrounding their use become more difficult. For example, the development of self-driving cars raises concerns regarding duty and culpability in the event of an accident. Furthermore, if not adequately regulated and monitored, the use of AI algorithms in decision-making processes such as the criminal justice system or loan applications may result in biases and discrimination [21]. Furthermore, Rainie and Anderson [22] add that technical developments can have far-reaching societal and economic consequences. With the rise of automation and AI-powered technology, there is growing anxiety about job displacement. Many traditional jobs may become obsolete as computers take over repetitive and routine labour, leading to unemployment and socioeconomic inequity. Furthermore, developing technologies have the potential to worsen the digital gap, with individuals without access to technology falling farther behind in terms of education, economic possibilities, and social inclusion.

The impact of developing technology on privacy is another big risk. The expansion of data-driven technologies, such as social media platforms and the Internet of Things, has generated worries about personal information privacy and security. Massive data gathering, storage, and analysis can result in information misuse, surveillance, and erosion of individual privacy rights. Regulation of the gathering and use of personal data, as well as data security promotion and individual knowledge, are crucial in limiting this danger [23]. Furthermore, there are environmental dangers associated with new technologies. For example, the creation, use, and disposal of electronic devices—like computers and smartphones—contributes to electronic trash, which presents serious risks to human health and the environment. Concerns over increased carbon emissions and their role in climate change are also raised by the rising energy demand needed to power these technological breakthroughs. To preserve the long-term viability of both our technological advancements and the environment, we must strike a balance between technological progress and sustainable practices [24].

Finally, there are concerns related to law and regulation that come with technical breakthroughs. Technology is developing at a rate that frequently surpasses the creation of suitable legal frameworks to control its application. In sectors like driverless vehicles, drones, and genetic engineering, in particular, this regulatory gap makes it difficult to define obligations and handle emergent hazards. It is crucial to strike the correct balance between promoting innovation and safeguarding the public interest in order to prevent moral and legal ambiguities [6].

4. Artificial intelligence-associated risks

Artificial intelligence (AI) has clearly achieved tremendous traction in recent years, revolutionising different industries and redefining the way we live and work. However, in addition to the numerous benefits provided by AI, there are substantial hazards and risks that must be handled.

One of the most serious threats of AI is job loss. Many functions currently performed by humans could be automated by AI technologies, raising fears about mass unemployment and economic disruption. As machines gain the ability to do sophisticated cognitive tasks, roles reliant on human decision-making and intelligence may become obsolete. Manufacturing, transportation, and customer service may be significantly affected. Finding a middle ground between AI adoption and job stability is a critical task for society [25]. According to Drage and Mackereth [26], AI's capacity for bias and discrimination represents a serious risk as well. Large volumes of data, including historical data that inevitably reflects societal biases, are fed into AI models so they may learn. Inequalities and prejudices already in place may be reinforced and amplified if AI systems are trained using this biassed data. AI algorithms that are employed, for instance, in recruiting procedures might unintentionally prejudice against particular demographic groups. In order to mitigate this danger, diverse and inclusive datasets must be promoted, and justice and openness must be ensured in the development and application of AI systems.

The creation and application of AI technology also raise ethical questions. This includes concerns about AI's decision-making and responsibility are getting more complicated as it develops. Autonomous cars, for example, need to be designed to make snap decisions in circumstances that could endanger lives. There are moral conundrums when deciding how an AI should weigh the value of human life against other considerations. These issues demand serious thought and analysis. To ensure ethically sound and responsible decision-making, strong ethical frameworks and norms are required to oversee the development and application of AI [21]. AI technologies also put security and privacy at risk. Since AI depends on massive datasets to work properly, there is a chance that personal data will be misused or accessed without authorization. This is especially problematic in light of the growing popularity of AI-powered applications like facial recognition, which are capable of gathering and analysing large volumes of sensitive data. Ensuring privacy is of utmost importance, and effective measures to reduce potential threats, including stronger data protection laws, encrypted communication, and improved cybersecurity, will be necessary [27]. Furthermore, Moisset [28] adds that there is a chance that AI will be misused for bad intentions. AI-driven cyberattacks have the capacity to be extremely destructive and smart. AI algorithms have the potential to produce convincing deepfakes, for instance, which may be used for impersonation or to disseminate false information. Moreover, social engineering attacks that target people or organisations in an effort to harm their reputations or obtain financial advantage can be automated and amplified with the help of AI. To mitigate these hazards, it is imperative to fortify cybersecurity protocols and increase cognizance regarding cyber threats associated with artificial intelligence.

Anderson [29] suggests that the possibility of unforeseen outcomes is another risk connected to AI. AI systems can unintentionally result in unanticipated consequences even though they are intended to optimise particular goals or address specific problems. For example, substantially biassed AI suggestions or decision-making may arise from biassed training data. AI algorithms used in stock market or financial trading may also inadvertently set off a chain reaction that destabilises the market. Mitigating these unexpected outcomes requires extensive risk assessments, continuous monitoring, and human oversight. Finally, risks associated with the creation of superintelligent AI must be carefully considered. Artificial intelligence (AI) systems that are capable of self-improvement and are smarter than humans are referred to as superintelligence. This could result in a sudden and significant rise in the systems' capabilities. This creates questions regarding the safety and control of such sophisticated AI, since it may be difficult for humans to understand and operate machines smarter than they are. Research and thought should be given to ensuring that superintelligent AI is in line with human ideals and that strong safety precautions are in place to avoid disastrous consequences [30].

5. Internet of things associated risks

Due to its ability to connect gadgets and facilitate previously unheard-of levels of data interchange, the Internet of Things, or IoT, has become an essential component of our everyday life. This network of connections has many advantages, but there are also major concerns that must be taken into consideration. Some of those risks are given below:

According to Tweneboah-Koduah et al. [31], security breach is one of the main dangers connected to the Internet of Things. The sheer number of internet-connected devices—billions—increases the risk of cyberattacks and illegal access to private information. Vulnerabilities in IoT devices can be exploited by hackers to obtain personal data or potentially take over vital infrastructure systems. Serious repercussions could result from this, including possible dangers to public safety and financial loss. Potential privacy invasion is another issue that comes with IoT. Large volumes of data, including sensitive and personal information, are frequently collected and transmitted by IoT devices. Using this information, comprehensive profiles of people's likes, behaviours, and habits can be constructed. Inappropriate use of this information may result in fraud, identity theft, and other types of discrimination [32]. Furthermore, IoT devices are vulnerable to security flaws because they lack uniform security measures. A lot of IoT devices sacrifice strong security features in order to be affordable and user-friendly. Because of this, hackers can easily target them and take advantage of their vulnerabilities to carry out nefarious operations. To safeguard user data, device makers must place a high priority on security by putting strong authentication, encryption, and access control systems in place [33].

Jang-Jaccard and Nepal [34] suggest that the Internet of Things raises safety issues in addition to security and privacy issues. Catastrophic failures are more likely to occur in networked vital systems, such as electricity grids, automobiles, and medical equipment. Via the manipulation of medical equipment, power supply systems, or traffic light controls, a compromised IoT device might cause havoc. This highlights how urgently strict safety guidelines and laws are needed to guarantee that Internet of Things devices are meticulously built and comply to quality control procedures. Das and Inuwa [35] add that the massive amount of data that linked devices generate and share is another problem associated with the Internet of Things. Real-time processing, analysis, and storage of this data are frequently required. But this presents problems with regard to system scalability, storage capacity, and data management. This deluge of data, if improperly managed, may cause network congestion, processing delays, and heightened susceptibility to intrusions. To effectively handle the increasing demand for IoT-generated data, businesses and service providers must invest in strong data infrastructure, including data centres, cloud platforms, and data analytics tools.

Moreover, contends Abderahman [36], during a product's whole lifecycle, the interconnectedness of IoT devices adds new hazards and complicates supply chains. Every stage of the process, from procuring components to deploying and disposing of devices, has possible weaknesses that could be used by bad actors. Strong supply chain management procedures are therefore required, together with stringent screening procedures, ongoing device monitoring, and third-party audits. Lastly, Hand [37] asserts that there are issues with the ethical ramifications of Internet of Things technology. There may be moral conundrums if massive volumes of personal data are gathered and used without the express agreement of the user. It is imperative to tackle concerns regarding transparency, informed consent, and responsible data usage to guarantee that Internet of Things implementations conform to moral standards and uphold the rights of persons.

6. Blockchain-associated risks

The potential of blockchain technology to transform a number of industries has drawn a lot of interest in recent years. Although blockchain has many advantages, like greater efficiency, security, and transparency, there are risks involved. Blockchain's vulnerability to cyberattacks is one of the main risks it carries. Despite the fact that blockchain networks are meant to be safe, there have been cases when weaknesses have been taken advantage of. For example, millions of dollars worth of cryptocurrencies were lost in the 2016 DAO (Decentralised Autonomous Organisation) breach. A security hole in the programming allowed the attacker to steal money. This emphasises how crucial it is to carry out exhaustive testing and code audits prior to using blockchain technology in order to reduce such dangers [38].

Habib et al. [39] potential privacy breaches are other risks connected to blockchain technology. Blockchain provides openness, but it also raises security concerns for sensitive data. All transactions and related data on a public blockchain are accessible to anybody with network connectivity. This raises questions about how private people's financial and personal data is. Utilising privacy-enhancing technology, like secure multi-party computation or zero-knowledge proofs, can help mitigate these dangers by preserving the advantages of blockchain transparency while allowing for data privacy. Furthermore, one major risk connected to blockchain technology is scalability. Transaction processing and validation times expand with the size and usage of blockchain networks. Scalability becomes especially important when thinking about implementing blockchain in applications that will be widely used, such as supply chain management or payment systems. Scalability issues have been addressed with a variety of solutions, including sharding and off-chain scaling techniques.

These solutions do, however, carry some dangers and difficulties of their own, such as diminished decentralisation or possible security flaws [40].

According to Singh [41], regulation-related risks are also connected to blockchain technology. Blockchain is still in its infancy and is being used by many businesses, which has left it in a regulatory ambivalence. Diverse legal frameworks have diverse stances on blockchain technology, which can be dangerous for companies and individuals involved in the industry. Establishing regulatory frameworks is necessary to guarantee adherence to know-your-customer (KYC) and anti-money laundering (AML) laws, as well as to promote industry innovation and expansion. The possibility of governance issues is another risk connected to blockchain technology. Due to the decentralised nature of blockchain networks, decisions about protocol modifications, network upgrades, and dispute resolution must be agreed upon by all parties involved. Due to varying stakeholder interests and points of view, this can be a difficult procedure. Hard forks, in which a blockchain divides into two distinct chains and maybe creates confusion and instability, might result from disagreements within the community. By guaranteeing that choices are made in a transparent, inclusive, and equitable manner, good governance tools and processes can help reduce this risk [42].

Clarke [43] asserts that concerns exist over the potential effects of blockchain technology on the environment. Verifying and appending transactions to the blockchain through mining necessitates a significant amount of computer power and energy. This has given rise to complaints that blockchain uses a lot of energy and worsens the environment and carbon emissions. Examining proof of stake (PoS) as an alternative to proof of work (PoW) for consensus can help lessen the environmental impact of blockchain technology. Finally, there is a risk related to blockchain's immutability. One of the primary characteristics of the blockchain is its immutability; nevertheless, this characteristic has a drawback as well. It becomes difficult to amend or erase false or fraudulent information once it is posted on the blockchain. For sectors where inaccurate information can have unfavourable effects, like supply chain management or healthcare, this could have major ramifications. These risks can be reduced by putting strong data validation and verification procedures in place and taking off-chain arbitration procedures into account for dispute resolution [44].

7. Biotechnology-associated risks

A new path for scientific progress and possible advantages has been made possible by biotechnology, which is the manipulation of living organisms to create or alter goods, procedures, or technologies. Still, there are risks and issues associated with it that need to be properly thought through and dealt with. Gene-modified organisms (GMOs) accidentally released into the environment is one of the main risks associated with biotechnology. GMOs are organisms whose genetic makeup has been modified through the application of contemporary biotechnology methods. Even while genetically modified organisms (GMOs) can have a lot of benefits, such as higher crop yields or stronger insect resistance, their unchecked release might endanger ecosystems and biodiversity. The delicate balance of our ecosystems may be irreversibly damaged by these modified organisms, which have the potential to outcompete native species, upend natural food chains, or introduce new illnesses [45]. Additionally, Caradus [46] suggests that there may be unanticipated health risks associated with biotechnology for both people and animals. Genetically engineered crops could, for instance, unintentionally cause allergic reactions or negatively impact the digestive system. Because of the complexity of genetic interactions, scientists test GM crops extensively before releasing them into the market, but unexpected repercussions can still occur. Furthermore,it's possible that altered genes from genetically altered organisms willaltered genes from genetically altered organisms may spread to unintended or wild species, with unknowable consequences for the environment and public health.

The improper use of genetically engineered organisms for detrimental ends is another possible biotechnology risk. Although biotechnology offers enormous promise for enhancing human welfare, there are worries that these potent instruments may be used for evil purposes. For example, the intentional development of genetically engineered organisms with the goal of developing bioweapons is a serious threat to international security. In order to implement strict restrictions and guarantee that biotechnology is utilised exclusively for peaceful, moral, and advantageous reasons, governments and regulatory authorities must collaborate [47].

Utilising biotechnology also raises ethical questions. While genetic modification holds promise for novel medical interventions and the prevention of disease, it also raises concerns about tampering with nature and the possibility of unintended consequences. For example, choosing particular genetic features to create "designer babies" raises ethical concerns regarding what constitutes appropriate human gene editing. Social cohesiveness may be compromised if it results in a community split along genetic lines and discrimination against due to genetic differences [48]. Furthermore, the livelihoods of farmers and conventional agriculture are at risk from biotechnology. The broad use of genetically modified crops may result in the concentration of agricultural power in the hands of a small number of companies. These companies regulate the seed supply and may bar small farmers who cannot afford their exorbitant costs by patenting genetically modified seeds. This concentration of power has the potential to worsen social inequality and lessen rural communities' variety in terms of both economy and culture [49]. The public's opinion of biotechnology carries certain risks as well. Notwithstanding the fact that scientific advancement in this area depends on research, resistance and opposition may arise from the general public's misinformation and fear of biotechnology. Protests and campaigns against genetically modified organisms (GMOs) have been sparked by public worries about GMOs. This has impeded research and development efforts and reduced the potential advantages that biotechnology may provide. A fair and knowledgeable discussion about biotechnology must take into account these worries and provide the public with correct information [50].

8. Limitations

The chapter may have a narrow scope, focusing on certain technological breakthroughs or industries, and hence may not provide a complete assessment of the consequences of risk management across diverse sectors. The chapter relied on theoretical frameworks or anecdotal evidence to support its claims rather than delivering empirical research. This lack of empirical evidence may have restricted the conclusions' validity and generalizability. The chapter may have not addressed all forms of risks related to technological breakthroughs. Certain developing risks or complexities may not have been fully addressed, limiting the paper's applicability for organisations operating in those sectors.

9. Recommendations

The tremendous technological breakthroughs of our day have opened up new frontiers, but they have also presented threats on a never-before-seen scale. As new technologies are adopted by organisations and individuals, it is critical to investigate the implications of risk management in this environment. The following section highlights the primary dangers linked with technology breakthroughs and proposes appropriate risk mitigation techniques.

9.1 Risks to cybersecurity

The rise in cybersecurity risks is one of the main effects of technological breakthroughs. Cyber threats have expanded in variety and sophistication as more devices become networked and valuable data is stored and transferred. It can be difficult for businesses and people to safeguard sensitive data against various security lapses like phishing, hacking, and data breaches. Various tactics can be employed to reduce these risks:

9.1.1 Robust encryption and verification

Securing data transmission and access control can be greatly improved by putting strong encryption and two-factor authentication into place. This is already happening in some organisations and businesses, but some still lagging behind.

9.1.2 Routine security evaluations

Regular security audits aid in finding weaknesses and providing appropriate solutions. It guarantees that security protocols are current and that new risks are dealt with right away.

9.1.3 Employee development

The chance of inadvertent data breaches can be reduced by developing employees about cybersecurity best practises. To increase public awareness of potential risks and how to mitigate them, regular training programmes ought to be held.

9.2 Privacy issues

Significant privacy concerns have also been generated by technological advancements.

The volume of personal data being gathered and handled makes it increasingly necessary to address the risks of privacy infringement. Organisations and individuals can take the following actions to reduce these risks:

9.2.1 Designing for privacy

Privacy breaches can be avoided by incorporating privacy protection at every step of system development. Ensuring that privacy is protected by default can be achieved through the use of privacy-enhancing technologies and privacy effect assessments.

9.2.2 Transparency and user consent

It is critical to provide individuals with clear information about how their data will be collected, utilised, and shared. Obtaining informed consent for data processing activities can aid in the development of trust and the reduction of privacy threats.

9.3 Regulation compliance

Compliance with privacy and data protection standards, such as the General Data Protection Regulation (GDPR) and in the case of South Africa the Protection of Personal Information Act (POPI Act), is critical in mitigating privacy threats. Organisations should verify that they are in compliance with applicable legislation and that they have systems in place to remedy any violations.

9.3.1 New technological risks

As technological improvements continue, new and unanticipated threats emerge. Artificial intelligence (AI) biases, autonomous car safety, and data ethics are some of the major hazards linked with developing technology. Organisations can consider the following ways to mitigate these risks:

9.3.2 Frameworks for ethics and responsibility

The creation and observance of ethical frameworks are essential for reducing the risks related to data ethics and AI biases. Organisations that develop and implement emerging technologies must have a responsible strategy that puts responsibility, fairness, and transparency first.

9.3.3 Strong testing and validation

Stringent testing and validation procedures must be in place for driverless vehicles and other new technology. Before extensive implementation, simulations, real-world testing, and continuous monitoring can help successfully identify and manage risks.

9.3.4 Collaboration and information sharing

It is critical to foster collaboration among regulators, industry professionals, and researchers in order to keep ahead of new dangers. Sharing knowledge, best practises, and lessons learnt can help organisations reduce the risks associated with technological breakthroughs collaboratively.

10. Conclusion

This chapter concludes that the complexity and scope of the risks that organisations confront have grown due to the rapid speed of technological improvements. For risk management to properly handle the possible impact on enterprises, technical risks and vulnerabilities must now be included as a distinct category. New risks, including cyber-attacks, data breaches, and technical obsolescence, have emerged in the age of technological breakthroughs. The dynamic nature of technology necessitates the updating and adaptation of risk management techniques. Risks associated with technology necessitate a proactive, dynamic approach to risk management that includes ongoing threat assessment and monitoring.

For organisations to effectively reduce and manage risks, they must invest in strong technology infrastructure and processes. Technological developments have given organisations new possibilities, but they also carry with them new risks that must be carefully evaluated and controlled. In the current era of technological progress, risk management must not be confined to a single department or procedure, but rather must be integrated into all levels and activities of an organisation. To effectively guide and help organisations, risk management professionals must be up to date on the newest technological innovations and the risks connected with them. Modern technology has eroded boundaries that were once clear and created new dangers from remote work, cloud computing, and mobile devices.

Regular evaluations of technology-related risks and vulnerabilities, along with the installation of suitable controls and safeguards, are essential components of risk management. As a result of technological developments, risk management is now a more analytical and data-driven discipline that requires sophisticated tools and methods to identify and mitigate risks. In the age of technological breakthroughs, organisations should take a proactive approach to risk management instead of depending just on reactive measures. The security of vital assets and systems must be given top priority in risk management due to the growing dependency of business operations on technology.

Risk management techniques must consider the possible legal and compliance challenges that come with technological progress, like laws pertaining to data security and privacy. The current technology era necessitates that organisations cultivate a culture of risk awareness and education so that all staff members are aware of the risks and can assist with risk management initiatives. With a common knowledge of the risks and their potential consequences, risk management in the age of technology breakthroughs involves cooperation and coordination between various departments and stakeholders.

The dangers associated with technological breakthroughs are both known and unknown, which emphasises the value of scenario planning and stress testing in risk management. Given the possible financial and operational effects of technological advancements, risk management should consider the quick obsolescence of technologies. To properly manage and lessen the impact of technology-driven risks, a mix of preventive, investigative, and remedial controls is frequently needed.

DOI: http://dx.doi.org/10.5772/intechopen.1003899

Author details

Monument Thulani Bongani Makhanya
University of Zululand, South Africa

*Address all correspondence to: makhanyamt@unizulu.ac.za

References

[1] Javaid M, Haleem A, Singh RP, Suman R, Gonzalez ES. Understanding the adoption of industry 4.0 technologies in improving environmental sustainability. Sustainable Operations and Computers. 2022;**3**:203-217. ISSN 2666-4127. DOI: 10.1016/j.susoc.2022.01.008

[2] Javaid M, Haleem A, Singh RP, Suman R. Substantial capabilities of robotics in enhancing industry 4.0 implementation. Cognitive Robotics. 2021;**1**:58-75. ISSN 2667-2413. DOI: 10.1016/j.cogr.2021.06.001

[3] Larmour John. The Future of Collaboration: How Technology is Changing the Way We Work Together. 2023. Available from: https://www.linkedin.com/pulse/future-collaboration-how-technology-changing-way-we-john-larmour-msc-/ [Acccessed: November 15, 2023]

[4] Gupta SP, Kushwaha S, Badhera U, Chatterjee P, Santibanez EDR, Gonzalez. Identification of benefits, challenges, and pathways in E-commerce industries: An integrated two-phase decision-making model. Sustainable Operations and Computers. 2023. ISSN 2666-4127. DOI: 10.1016/j.susoc.2023.08.005

[5] Haleem AM, Javaid MA, Qadri RS. Understanding the role of digital technologies in education: A review. Sustainable Operations and Computers. 2022;**3**:275-285, ISSN 2666-4127. DOI: 10.1016/j.susoc.2022.05.004

[6] Fenwick Mark D, Kaal Wulf A, Vermeulen Erik PM. Regulation tomorrow: What happens when technology is faster than the law? American University Business Law Review. 2017;**6**(3):561-594. Available from: http://digitalcommons.wcl.american.edu/aublr/vol6/iss3/1

[7] Karin Kelley. What is cybersecurity and why it is important? 2023. Available from: https://www.simplilearn.com/tutorials/cyber-security-tutorial/what-is-cyber-security [Accessed: November 15, 2023]

[8] Tariq U, Ahmed I, Bashir AK, Shaukat K. A critical cybersecurity analysis and future research directions for the internet of things: A comprehensive review. Sensors. 2023;**23**(8):4117. DOI: 10.3390/s23084117

[9] Kavanagh Camino. New tech, new threats and governance challenges: An opportunity to craft smarter responses? 2019. Available from: https://carnegieendowment.org/2019/08/28/new-tech-new-threats-and-new-governance-challenges-opportunity-to-craft-smarter-responses-pub-79736 [Accessed: November 15, 2023]

[10] Organisation Economic Corporation and Development. Regulatory effectiveness in the era of digitalisation. 2019. Available from: https://www.oecd.org/gov/regulatory-policy/Regulatory-effectiveness-in-the-era-of-digitalisation.pdf [Accessed: November 15, 2023]

[11] Farida I, Setiawan D. Business strategies and competitive advantage: The role of performance and innovation. Journal of Open Innovation: Technology Market, and Complexity. 2022;**8**(3):163. DOI: 10.3390/joitmc8030163

[12] Ritesh Jain, Fritz Nauck, Thomas Poppensieker, Olivia White. Meeting the future: Dynamic risk management for uncertain times. 2019. Available from: https://www.mckinsey.com/

capabilities/risk-and-resilience/our-insights/meeting-the-future-dynamic-risk-management-for-uncertain-times [Accessed: November 15, 2023]

[13] Taherdoost H. Data collection methods and tools for research; a step-by-step guide to choose data collection technique for academic and business research projects. International Journal of Academic Research in Management (IJARM). 2021;**10**(1):10-38. ffhal-03741847f

[14] Goundar Sam. Chapter 3—Research Methodology and Research Method. 2012. Available from: https://www.researchgate.net/publication/333015026_Chapter_3_-_Research_Methodology_and_Research_Method [Accessed: November 15, 2023]

[15] Johnson Melissa P. Secondary data analysis: A method of which the time has come. Qualitative and Quantitative Methods in Libraries (QQML). 2014;**3**:619-626, ISSN 2241-1925

[16] Kabir Syed Muhammad. Methods of data collection, chapter: 9. In book: Basic Guidelines for Research: An Introductory Approach for all Disciplines. vol. 1. Chittagong, Bangladesh: Book Zone Publication; 2016

[17] Bowen G. Document analysis as a qualitative research method. Qualitative Research Journal. 2009;**9**:27-40. DOI: 10.3316/QRJ0902027

[18] Ahn E, Kang H. Introduction to systematic review and meta-analysis. Korean Journal of Anesthesiology. 2018;**71**(2):103-112. DOI: 10.4097/kjae.2018.71.2.103. Epub 2018 Apr 2

[19] Păvăloaia V-D, Necula S-C. Artificial intelligence as a disruptive technology—A systematic literature review. Electronics. 2023;**12**(5):1102. DOI: 10.3390/electronics12051102

[20] Krause T, Ernst R, Klaer B, Hacker I, Henze M. Cybersecurity in power grids: Challenges and opportunities. Sensors (Basel). 2021;**21**(18):6225. DOI: 10.3390/s21186225

[21] Eleanor B, Jasmin F-S, Nicola J, Ruth L, Emma W, Winfield A. The Ethics of Artificial Intelligence: Issues and Initiatives. Brussels: European Union; 2020. PE 634.452. DOI: 10.2861/6644. ISBN: 978-92-846-5799-5

[22] Lee R, Janna A. The Future of Jobs and Jobs Training. Pew Research Centre; 2017. Available from: https://www.pewresearch.org/internet/2017/05/03/the-future-of-jobs-and-jobs-training/ [Accessed: November 16, 2023]

[23] Sara Q, Park T, Weaven KD, Martin WS, Palmatier RW. Digital technologies: Tensions in privacy and data. Journal of the Academy of Marketing Science. 2022;**50**:1299. DOI: 1323, 10.1007/s11747-022-00845-y

[24] Geneva Environmental Network. The Growing Environmental Risks of E-Waste. Geneva: Geneva Environmental Network; 2023. Available from: https://www.genevaenvironmentnetwork.org/resources/updates/the-growing-environmental-risks-of-e-waste/ [Accessed: November 16, 2023]

[25] Frank MR, Autor D, Bessen JE, Brynjolfsson E, Cebrian M, Deming DJ, et al. Toward understanding the impact of artificial intelligence on labor. Proceedings of the National Academy of Sciences. 2019;**116**(14):6531-6539. DOI: 10.1073/pnas.1900949116

[26] Drage E, Kerry M. Does AI debias recruitment? Race, gender, and AI's "eradication of difference". Philosophy Technology. 2022;**35**(4):89. DOI: 10.1007/s13347-022-00543-1

[27] Khalid N, Qayyum A, Bilal M, Al-Fuqaha A, Qadir J. Privacy-preserving

artificial intelligence in healthcare: Techniques and applications. Computers in Biology and Medicine. 2023;**158**:106848, ISSN 0010-4825. DOI: 10.1016/j.compbiomed.2023.106848

[28] Moisset Sonya. The effects of AI in cybersecurity handbook—The malicious use of AI in cyberattacks. 2023. Available from: https://www.freecodecamp.org/news/effects-of-ai-in-cybersecurity-handbook/ [November 16, 2023]

[29] Lindsey A. Artificial intelligence in international development: Avoiding ethical pitfalls. Journal of Public Affairs and International Affairs, Princeton University. 2019:1. Available from: https://jpia.princeton.edu/news/artificial-intelligence-international-development-avoiding-ethical-pitfalls [Accessed: November 16, 2023]

[30] Ebhardt JD. The threat of artificial superintelligence. Exigence. 2018;**2**(1):1-8. Available from: https://commons.vccs.edu/exigence/vol2/iss1/1 [Accessed: November 16, 2023]

[31] Tweneboah-Koduah S, Skouby KE, Tadayoni R. Cyber security threats to IoT applications and service domains. Wireless Personal Communications. 2017;**95**:169-185. DOI: 10.1007/s11277-017-4434-6

[32] Mao C. Privacy Issues in IoT: Privacy Concerns in Smart Home [Thesis]. Linnaeus: Linnaeus University; 2019

[33] Tariq U, Ahmed I, Bashir AK, Shaukat K. A critical cybersecurity analysis and future research directions for the internet of things: A comprehensive review. Sensors. 2023;**23**(8):4117. DOI: 10.3390/s23084117

[34] Julian J-J, Nepal S. A survey of emerging threats in cybersecurity. Journal of Computer and System Sciences. 2014;**80**(5):973-993, ISSN 0022-0000. DOI: 10.1016/j.jcss.2014.02.005

[35] Resul Das, Inuwa Muhammad. A review on fog computing: Issues, characteristics, challenges, and potential applications. Telematics and Informatics Reports. 2023;**10**:100049. ISSN 2772-5030. DOI: 10.1016/j.teler.2023.100049

[36] Rejeb A, Suhaiza Z, Rejeb K, Seuring S, Treiblmaier H. The internet of things and the circular economy: A systematic literature review and research agenda. Journal of Cleaner Production. 2022;**350**:131439, ISSN 0959-6526. DOI: 10.1016/j.jclepro.2022.131439

[37] Hand DJ. Aspects of data ethics in a changing world: Where are we now? Big Data. 2018;**6**(3):176-190. DOI: 10.1089/big.2018.0083

[38] Rasool MA, Shafiq HM. Blockchain Technology: A New Domain for Cyber Forensics [Thesis]. Sweden: Halmstad University; 2018

[39] Habib G, Sharma S, Ibrahim S, Ahmad I, Qureshi S, Ishfaq M. Blockchain technology: Benefits, challenges, applications, and integration of blockchain technology with cloud computing. Future Internet. 2022;**14**(11):341. DOI: 10.3390/fi14110341

[40] Rožman N, Corn M, Škulj G, Berlec T, Diaci J, Podržaj P. Exploring the effects of blockchain scalability limitations on performance and user behavior in blockchain-based shared manufacturing systems: An experimental approach. Applied Sciences. 2023;**13**(7):4251. DOI: 10.3390/app13074251

[41] Yogender S. The Legal and Regulatory Landscape of Blockchain Technology. Orlando: IGI Global; 2023.

Available from: https://www.linkein.com/pulse/legal-regulatory-landscape-blockchain-technology-yogender-singh/ [Accessed: November 16, 2023]

[42] Tan E, Mahula S, Crompvoets J. Blockchain governance in the public sector: A conceptual framework for public management. Government Information Quarterly. 2022;**39**(1):101625, ISSN 0740-624X. DOI: 10.1016/j.giq.2021.101625

[43] Clarke Anthony. The Environmental Impact of Blockchain Technology. 2023. Available from: https://www.nasdaq.com/articles/the-environmental-impact-of-blockchain-technology [Accessed: November 16, 2023]

[44] Sharif MM, Ghodoosi F. The ethics of blockchain in organizations. Journal of Business Ethics. 2022;**178**:1009-1025. DOI: 10.1007/s10551-022-05058-5

[45] Arcieri M. Spread and potential risks of genetically modified organisms. Agriculture and Agricultural Science Procedia. 2016;**8**:552-559. DOI: 10.1016/j.aaspro.2016.02.072

[46] Caradus JR. Intended and unintended consequences of genetically modified crops—Myth, fact and/or manageable outcomes? New Zealand Journal of Agricultural Research. 2023;**66**(6):519-619. DOI: 10.1080/00288233.2022.2141273

[47] John E, Robert G. Feeding the World: How Changes in Biotech Regulation Can Jump Start the Second Green Revolution and Diversify the Agricultural Industry. Vol. 327. Virginia: 44 William & Mary Environmental Law and Policy Review; 2020. Arizona Legal Studies Discussion Paper No. 19-17. Available from: https://ssrn.com/abstract=3437520 or 10.2139/ssrn.3437520

[48] Lau Pin Lean. Evolved eugenics and reinforcement of "othering": Renewed ethico-legal perspectives of genome editing in reproduction. BioTechnology (Basel). 2023;**12**(3):51. DOI: 10.3390/biotech12030051

[49] National Academies of Sciences, Engineering, and Medicine; Division on Earth and Life Studies; Board on Agriculture and Natural Resources; Committee on Genetically Engineered Crops: Past Experience and Future Prospects. Genetically Engineered Crops: Experiences and Prospects. Vol. 6. Washington (DC): National Academies Press (US); 2016, Social and Economic Effects of Genetically Engineered Crops. Available from: https://www.ncbi.nlm.nih.gov/books/NBK424536/

[50] Raghavan AK, Bawa AS. Genetically modified foods: Safety, risks and public concerns—A review. Journal of Food Science and Technology. 2013;**50**(6):1035-1046. DOI: 10.1007/s13197-012-0899-1

Section 2

Public and Ergonomic Approach

Chapter 9

Public Management of Risks and Threats in the Conditions of the Martial Law of the European State

Valerii Ye. Vorotin

Abstract

The article substantiates the approaches to solving the actual scientific problem of the theory and practice of public administration – the peculiarities of public (state and local) administration in the conditions of security violations and martial law on the territory of the country. Such an analysis was carried out on the basis of a clear definition of the place and role of the anti-crisis management system in the balance of social relations in the public sphere and the development of mechanisms for coordinating the actions of management subjects. The content of the main concepts of the anti-crisis public management is defined. The methodology of the stages of development of public administration has been developed, which is aimed at a meaningful classification – the expediency of using the mechanisms and tools of anti-crisis management as a means of achieving the goal in the conditions of martial law and post-war recovery of the European state. The essence of individual mechanisms and tools for ensuring management in the field of risks to human life is revealed. The information scheme of the functioning of the institutional mechanism of anti-crisis management in the conditions of martial law in a separate European state has been developed.

Keywords: public administration, regional administration, risk, security, resource support for development, mechanisms of state administration, martial law

1. Introduction

The main tasks facing the Ukrainian state today are preserving sovereignty; ensuring economic, resource, civil, and food security; accelerating European and Euro-Atlantic integration; restoring sustainable and dynamic economic development; restoring a decent standard of living and security of citizens; increasing the effectiveness of the state administration system, and regional and local development under martial law. The resolution of these and other issues largely depends on the effective legal regulation of the functioning of the mechanisms and tools of state administration as a whole and its basis – ensuring the management of the development of territories, especially in the sphere of life support.

In the conditions of martial law, there is a negative trend of increasing the aggressive influence of an external threat (the war of the Russian Federation against Ukraine) on the system of public administration in Ukraine, in particular, on the effectiveness of the functioning of the organizational and legal mechanisms of resource provision for economic development, which were practically not adapted to the conditions of martial law in the state and needed prompt regulatory settlement and adjustment. This work is carried out by the Verkhovna Rada of Ukraine 24/7. A list of laws and resolutions are adopted under martial law [1].

We note that risks and threats to the entire system of management and economic (resource) security in particular emphasize the assessment of vulnerable component mechanisms and tools of the system of public (state and local) management and administration in the conditions of martial law in Ukraine. We believe that these components are significant structural disparities, monopoly and closure of certain industries, obstacles to the development of the military and defense sectors, underestimation of civil protection of the population and territories, unresolved energy and technological dependence of the economy on the Russian Federation, uncontrolled outflow of material and financial resources outside of Ukraine, activity of shadow structures, presence of enemy agents of the Russian Federation, absence of a professional school of public management and administration, and others.

It is clear that, today, the domestic system of public management of national, regional, and local development needs latest approaches to the reality of Russian military aggression on the territory of Ukraine, which in turn actualizes the research, development, and introduction into domestic practice of foreign experience (initiatives) based on modern management approaches in the conditions of martial law, especially with regard to the development of communities in terms of resource livelihoods.

We agree with the opinion of a well-known expert that the Ukrainians managed to defeat and oust the Russians from the territories they have now captured. There are several reasons for this, – writes the American philosopher Francis Fukuyama in his column for americanpurpose.com, where he analyzes the causes of aggression and the path to the destruction of the Russian Federation. As early as March 10, 2022, F. Fukuyama published the article "Preparation for Defeat," in which he claims that Russia may be moving toward complete defeat in its war in Ukraine. At the time, the material attracted considerable attention in the world, and many people decided that the author was very optimistic. Most of this prediction was confirmed today. In fact, the Russians were defeated in their attempts to conquer Kyiv and retreated from northern Ukraine in the beginning of April.

Today, in the socio-economic literature, there is no single system of classification of risks and threats of managerial activity, but on the contrary, there are many approaches that, for the most part, determine only the purpose and individual tasks of management. Taking into account the shortcomings of the already developed classifications of risks, as well as analyzing the peculiarities of the methodology of developing classifications, we consider it necessary to carry out an analysis of the system of public management of risks and threats in the conditions of martial law in Ukraine.

2. Public administration: theory and risks of use

The field of science-public management is formed from a set of economic, legal phenomena and social relations in society, which characterize the sphere of interaction between the state and man (society) as an object of management and self-management at the local level.

The main thing that makes up the content of public administration is the most important functions of the state as an institution of public administration, the interaction of regularities (laws) and the principles of their implementation, scientific justification of the need for state policies as the basis of management, and implementation of local self-government. Functional analysis of the state's management activities, the organization of government bodies as a political and legal institution of the managing society, and the corresponding organizational, legal, and social anti-crisis relations formed at all management levels, is a subject of public management, in particular, in the conditions of martial law in the state.

In general, management is goal setting (conscious, foreseen, and thought out), organization and regulation of human interaction related to one's own social, collective, and group life activities, which is carried out both directly (in forms of self-management) and through specially created institutions (institutes).

Modern public administration relies on state power and is supported and ensured by it in terms of achieving goals and anti-crisis development of the entire economic system. Public management extends to the entire society in the direction of state restrictions on activity (management) and local development, in particular during martial law and the subsequent restoration of territories. It is the state through legislative activity that establishes basic, general, and typical rules (norms) of people's behavior. This is especially important in the conditions of martial law.

The state provides public management with regulation mechanisms and tools. Unlike other types of management, public management cannot become real without a system and mechanisms for implementing specific policies. We believe that public administration differs from state administration in that it uses all the possibilities of the rule of law, thanks to the use of existing mechanisms, laws, and control of society over all bodies of state power and effective local self-government.

Management as a specific phenomenon begins when there is a conscious public principle, interest and knowledge, goals and will, energy and human activity in any interconnections, relationships, and .processes.

Public administration is formed by people for conscious self-regulation of their life activities and has the same importance for ensuring their needs and interests as family and property, morality and law, production method and the state, knowledge and information, and so on. All these determine the dependence of management on the state of society, ideals, and values.

At different historical moments, management is reproduced by the corresponding society, "takes" its essence from it, realizes its capabilities, and is characterized and developed according to the level of its development.

Public administration is one of the most complex and responsible spheres of intellectual and practical activity of people and society. This is an area on which the well-being of the country and, ultimately, the fate of every person depends to a large extent. In management, everything comes from the person and is oriented toward him [2].

That is why, public administration in society arose, which is always relevant, as society is an open system because nothing in it is given once and for all, is not unchanging, and absolutely stable. Personal and social relationships are reproduced daily. Without it, chaos, anarchy, and decay ensue. In conditions of openness, management serves as the most important social institution of self-preservation and self-renewal.

We believe that it is not entirely correct to define management as an action or only influence in society. Its more adequate concept is the category of "relationships," which defines the main thing in management because it influences the consciousness, behavior, activity of people, and their mutual relations.

3. A new stage in the formation of anti-war state administration

The martial state of our country's development is complicated by the objective necessity of the urgent implementation of a set of urgent measures in all spheres of Ukrainian society, in particular, state (state and local self-government) administration. Today, the dominant role is occupied by the task of scientific substantiation of the functioning (adaptation) of the integral system of state administration in the conditions of martial law, and its priority is the system of personnel life support.

Research positions of scientists regarding the theoretical understanding of "public administration" have different interpretations and completely polar views. Some authors interpret public management as the influence of a separate subject of management, endowed with public power, on a certain object for the purpose of any results of social relations. Others, such as rule-making, administrative, economic-control, organizational-legal, and other activities, are carried out on the basis of developed regulations, rules, and procedures. The third is how to manage a specific situation together with relationship partners.

Let us consider the evolution of state administration with the traditional separation of separate periods of time (stages), which specify the mechanisms and tools of a specific model of administration that is used in this period in the state.

Stage 1 (from 1880 to 1920) is the initial study of state management.

The time of the initial stage 1, the study of public administration itself, was 40 years. Famous researchers of this stage are V. Wilson, M. Weber, A. Fayol, and others. The essence of public administration as a separate scientific direction was first investigated by V. Wilson. His scientific research proved that there are differences between other sciences, including political science, law, and the science of public administration. Thus, V. Wilson noted that the purpose of this science is the study of specific activities and directions of influence of the authorities, as well as managerial actions to carry out this work with minimal costs and as efficiently as possible. His research was aimed at identifying and analyzing the principles of public administration. He attributed to them the separation of political and management issues, comparative analysis of business and policy structures, research of the civil service, and improvement of its efficiency by using the experience of managing business structures in the work of state bodies.

The functioning of all public administration systems is carried out in such a way that every civil servant who is empowered and has a certain competence and responsibility, regardless of whether he is a high-ranking manager or a junior specialist in the ministry, has the power of decision and establishes the universal, public focus of this management decision. Therefore, the decision made on the spot should remain the main and predominant tool in the field of management, which combines the political and managerial vector of the development of social relations in the state. In our opinion, it is important that such relations are the object of state administration and reveal their social significance. Our analysis of the first stage of the development of public administration proved that it is based on a comparative analysis of the contradiction "politics/management," and research in this area made it possible to reveal the interdependence of the development of politics and management.

Stage 2 took place (from 1920 to 1950) the theory of management at the microlevel or individual economic entity. Economic relations in the space of contacts of private and public economic sectors were investigated. This stage took place at the level of 30 years. R. Simon, D. Waldo, M. Follet, E. Mayo, A. Maslow, and others became prominent representatives of this period.

At this stage, an interesting school of "human relations" developed. M. Follet, E. Mayo, and A. Maslow became the main researchers in this direction. The researchers focused their own research on the behavior of individuals and their activities in the organization.

At this stage, a general theory of management was formulated, which exists within the framework of the interaction of the private and public sectors, in particular, the principles of enterprise management, which have significant differences from state management, were studied.

Stage 3 (from 1950 to 1990) is a study of state policies and necessary management in a separate sphere of social relations. At this stage of the development of the theory of public administration, the works of P. Aucoin, F. Thompson, D. Truman, K. Hood, and others are of greatest interest. We believe that the researchers used effective directions, methods, and tools that were borrowed from the nonstate sector of the economy (real competition, hiring and contract systems, approaches to making the management decision itself, and the formation and implementation of state policies by increasing the efficiency of free decision-making). Mechanisms of adaptive ability to respond to a constantly changing environment, formation of orientation to developing strategy and policy, etc., were also investigated.

During this period, the concept of "new public management" was formed in the 1980s and was realized by a combination of market management mechanisms and management ideas and technologies of business entities of the private sector of the economy. According to this concept, executive authorities are perceived as economic entities that achieve results and provide management services to the population.

Also, the analysis of stage 3 proved that there was a real vector and separate tendencies to research the essence of modern public administration using socio-psychological and behavioral approaches, as well as the use of market laws and principles in the system and mechanisms of public administration.

Stage 4 (from 1990 to 2000) is the direction of "renewed management." For 10 years, the research stage of public administration continued. The following scientists and practitioners were engaged in this: T. Gebler, D. Osborn, P. Plastrik and others. Scientists studied state administration itself from the point of view of the effective conditions of business entities.

It is important that such an "economic approach" to the functioning of the management system in the state involves the use of market laws for the effective functioning of bureaucratic structures in order to increase the efficiency of their work and achieve the best indicators in the ratio of costs and results of activities in the field of public administration.

The study of this stage of the formation of public administration finds ways of fundamental changes in the understanding of the goals of public administration and the transition from minimum requirements to the maximum result of the activity of the state administration system, which, in addition to solving the main tasks, provides for the performance of many additional functions, particularly aimed at improving the quality of human life.

Stage 5 (from 2000 to 2022) is the direction of "increasing the effectiveness of public administration." The beginning of the XXI century laid the foundation for research in the field of public administration, which continued until the beginning of the Russian-Ukrainian war in Europe. Representatives of this direction were R. Brittany, M. Kartel, L. Jacobson, and others. We believe that, in this period, directions and mechanisms for improving the efficiency of public authorities and the quality of administrative services provided by the state are being investigated in the context of the development of democracy and society, which receive priority and special social significance. During this stage, there were studies of modern communication, which mainly characterized the interaction between representatives of state and private structures. In this regard, such research contributed to the adoption of socially significant decisions and made it possible to implement the functions of reforming the public administration system based on European management practice.

Within this stage, the concept of efficiency is often used in connection with the concepts of effectiveness, productivity, and efficiency of functioning. As for public administration, efficiency is usually associated with the achievement of the goals of public authority, the completeness and quality of the state's performance of its main functions.

In addition, here is stage 6 (from February 24, 2022, to the present) which the direction of anti-crisis or military public administration and testing of the system under the conditions of martial law in Ukraine. The beginning of 2022 is marked by the beginning of Russia's invasion of Ukraine and the beginning of a war in the center of Europe. Perhaps, this stage is still in its infancy, but it is important from the point of view of public administration in the sphere of protecting sovereignty and democratic values around the world [3].

Thus, the analysis of the modern stage of public administration research showed the relevance of the formation of a new concept of public administration, which is based on the public and especially the anti-war character of government in the state.

4. Economic security of the state, stimulation, and support of public administration

The main tasks facing the Ukrainian state today are preserving sovereignty; ensuring economic, resource, civil, and food security; accelerating European and Euro-Atlantic integration; restoring sustainable and dynamic economic development; restoring a decent standard of living and security of citizens; and increasing the effectiveness of the public administration system (state, regional, and local) under martial law. The resolution of these and other issues largely depends on the effective legal regulation of the functioning of the mechanisms and tools of public administration as a whole and its basis – ensuring the management of the development of territories, especially in the sphere of life support.

In the conditions of martial law, there is a negative trend of increasing the aggressive influence of an external threat (the war of the Russian Federation against Ukraine) on the system of public administration in Ukraine, in particular, on the effectiveness of the functioning of the organizational and legal mechanisms of resource provision for economic development, which were practically not adapted to the conditions of martial law in the state and needed prompt regulatory settlement and adjustment. This work is carried out by the Verkhovna Rada of Ukraine 24/7.

It should be noted that risks and threats to the entire management system emphasize the assessment of vulnerable component mechanisms and tools of the system of public management and administration under the conditions of martial law in

Ukraine. We believe that these components are significant structural disparities, monopoly and closure of certain industries, obstacles to the development of the military and defense sectors, underestimation of civil protection of the population and territories, unresolved energy and technological dependence of the economy on the Russian Federation, uncontrolled outflow of material and financial resources outside of Ukraine, the activity of shadow structures, the presence of enemy agents of the Russian Federation, the absence of a professional school of public management and administration, and others.

It is clear that, today, the domestic system of public management of national, regional, and local development needs the latest approaches to the reality of Russian military aggression on the territory of Ukraine, which in turn actualizes the research, development, and introduction into domestic practice of foreign experience (initiatives) based on modern management approaches in the conditions of martial law.

Russia's military aggression against Ukraine is unprecedented in terms of duration, scale, and devastating economic consequences. The Iran-Iraq war of the 1980s lasted almost 8 years and cost the Iraqi economy $452.6 billion (seven GDPs of the country at that time). Russia's five-day military aggression against Georgia in 2008 cost the latter about 23% of GDP (2 billion euros). Japan lost a quarter of its national wealth and a third of its production capacity in World War II [4].

According to preliminary estimates by experts, the economy of Ukraine lost up to 50% of the production capacities of enterprises in strategic industries, most of which were located in the east of Ukraine. The result of active hostilities was the disruption of both external and internal economic relations, the destruction of the country's housing stock, the destruction of production, social, and service infrastructure, and the blocking of seaports and other logistics channels. More than 80% of Ukrainian enterprises have reduced or stopped their activities. The fall in economic activity caused a reduction in tax revenues to the budget, limiting the government's ability to implement independent recovery policies.

According to the estimates of the Ministry of Economy of Ukraine, the gross domestic product of our country will decrease by 30.4% in 2022, which, although it is the largest drop in the recent history of Ukraine, is significantly less than the forecast level of international financial organizations (45–50%).

Critical to the country's further recovery and development are humanitarian risks associated with the loss of human capital and redistribution of labor within the country and loss of household income and livelihoods. According to United Nations estimates, as of January 17, 2023, the number of Ukrainian refugees registered for temporary protection in European countries reached 4.9 million people, of whom only a third expressed hope to return.

As the beginning of the full-scale Russian invasion, almost 4.8 million internally displaced persons have been officially registered in Ukraine, the increase of this socially vulnerable category of citizens causes significant pressure on the system of administration of public services.

Unemployment in the country increased sharply (from 9.9 to 30% according to the ILO methodology), adding to the imbalance of regional labor markets.

The Ukrainian state was able to avoid the most negative and pessimistic scenarios not only thanks to the unprecedented support of partner countries, which, according to the estimates of the National Bank of Ukraine, reached 120 billion dollars in 2022, of which 28 billion dollars was directed to finance the needs of the budget of Ukraine, but also due to the capacity of the public administration system, the stability of the institutions built over the years of systemic reforms.

There are the following systemic factors that ensured the effective functioning of public administration mechanisms and tools during the war:

- Developed economic system – entrepreneurial initiative, resource-based and network interaction, and lack of hierarchical centralized management of processes of redistribution of public resources made it possible to respond promptly to the challenges of wartime.

- Developed communications – a flexible system of transport, trade, information, and telecommunications ties preserved the "connectivity" of the country, mitigated the humanitarian crisis, and facilitated the coordination of actions of state and public institutions.

- Decentralization of public power – the broad powers of competent communities provided an opportunity to strengthen the defense capacity of the state.

- A developed civil society – the volunteer movement and a high level of self-organization of society became the basis of the stability of basic social systems.

- High level of openness of public administration – economic, political, and civil communications with foreign partners established in previous years made it possible to quickly attract help from the international community in financial (grants and loans), economic (material aid), humanitarian (humanitarian goods, programs support of refugees, and treatment of servicemen), and military (provision of weapons, ammunition, ammunition, and training of servicemen) spheres.

- Digitalization – the level of digitalization of the spheres of state administration achieved before the start of the full-scale war enabled the effectiveness of public authorities and the performance of duties vital for the functioning of the country.

The institutional nature of such factors makes it possible to consider them as characteristic features of the public administration system that has developed in Ukraine as a result of the entire previous period of social changes. In particular, due to Russia's full-scale military aggression in Ukraine, the International Monetary Fund has worsened the forecast for the growth of the world economy. The IMF expects the world economy to grow by 3.6% in 2022 and 2023. This is a sharp slowdown compared to 6.1% growth in 2021.

It should be noted that the macroeconomic trends of the state's development (2024–2030), which are in the forecasts of the Ukrainian government and international financial organizations are still not very positive. The probability of Ukraine's victory in the war against Russia is the end of 2024 - the beginning of 2025 [5].

The Ukrainian economy reached the bottom of its decline in April–May 2022, and since then, despite any intensity of shelling of its territory, it has shown a slow but stable trend of recovery of the main macroeconomic indicators. Thus, Ukraine's GDP in US dollars, which fell from $200 to $158 billion in the first year of the war, will rise to $170 billion in the second year, which will be the starting point for dynamic recovery. Investment, employment, savings, foreign and domestic trade show identical dynamics, improving steadily since summer 2022, despite ongoing large-scale hostilities. And we, as before, observe this improvement in 2023 as well. Absolutely all macroeconomic indicators show that Ukraine has already passed the bottom of its economic crisis related to the war.

At a conference in the Swiss city of Lugano, Ukraine announced the necessary amount of investment for recovery and received a positive response from representatives of the USA, the EU, and international financial organizations regarding potential sources of funding. This amount is ~$750 billion, of which ~200–300 billion is planned to be raised through partnership grants and borrowings in 2023–2025, and in 2026–2032, another $50 billion in 2024–2025 and $200 billion are expected in the form of private investments. In the first 3 years after the end of the war, partner aid is mainly expected, and in 2026–2032, on the contrary, the focus shifts to private investments. Briefly summarizing all the factors listed above, he considers a highly probable scenario in which Ukraine will "catch up" with its pre-war economic development trajectory by 2030. At the same time, the country will maintain double-digit growth figures for the next eight post-war years. By 2030, the indicative GDP figures are expected to be $440 billion, and the average salary is $1150.

5. Directions of ensuring management in the field of human life risks

Accidents, catastrophes, which are accompanied by deaths, injuries, shortening of life expectancy, and damage to health and the natural environment are the consequences of the manifestation of hazards, in particular, the management system. There is always the problem of assessing these consequences. The quantification of damages caused by the danger depends on many factors, for example, the number of people who were in the danger zone, the amount, and quality of material values that were there, natural resources, etc.

Each type of damage has its own quantitative expression. For example, the number of dead, wounded or sick, the area of the infected territory, the area of the forest that burned, the value of the destroyed buildings, etc.

The first quantitative method of determining damage is value-based, that is, determining damage in monetary terms. The second, universal, most widespread assessment of danger is risk. It can also be called a factor of potential danger. The concept of "risk" is "perceived possibility of danger." Obviously, the definition "perceived probability of danger" should also be considered more accurate.

Risk is a quantitative assessment of danger. It is the ratio of the number of certain undesirable realized consequences (n) to the maximum possible number of them (N1) for a specific period of time: $R = n/N$, that is, it is the frequency of realization of dangers. It is a companion of any human activity.

The above formula allows you to calculate the dimensions of an individual, group, and general risk. When assessing the general risk, the value N determines the maximum number of all events, and when assessing the group risk, the maximum number of events in a specific social group was selected from the total number based on a certain characteristic. The group can include people belonging to the same profession, age, and gender; the group can consist of one class of economic activity subjects [6].

According to the degree of admissibility for society, the risk is neglected, acceptable, marginally permissible, and excessive. Negligible risk has such a small level that it is within the permissible deviations of the natural level. Acceptable is the level of risk that the public can calmly accept, taking into account the technical, economic and social possibilities at this stage of the development of civilization. The maximum acceptable risk is the largest risk that cannot be exceeded, regardless of the expected result. Excessive risk is characterized by an extremely high level, which in the vast majority of cases leads to negative consequences.

Today, it is practically impossible to achieve a zero level of risk, that is, absolute safety. Absolute security cannot be guaranteed to any person regardless of their lifestyle and social status. We live because we avoid danger every day.

The demand for absolute security, attractive for its humanity, can turn into a tragedy for people. At this stage of society's development, the neglected risk is also impossible to achieve given the technical and economic prerequisites for this. Therefore, the modern concept of life safety is based on achieving an acceptable risk. The essence of the idea of the concept of "tolerable risk" is to try to create such a low level of risk that society perceives at the moment. Tolerable risk combines technical, economic, social, and political aspects and is a certain compromise between the level of security and the possibilities of its achievement.

The analysis of scientists' approaches to the definition of the concept of "risk" allows us to form signs by which its meaning is perceived. According to the first sign – possibility and probability of loss of resources – risk is characterized by possible success or loss of part of resources, losses, and lack of income in alternative developments of certain management actions. That is, in this aspect, risk is considered the result of an action and its consequence, which is determined by the influence of uncertainty factors.

J.M. Keynes generally interprets risk as a part of the value associated with possible losses caused by an unforeseen change in market prices, excessive operation of equipment, or destructive consequences of disasters.

In our opinion, risk is an objective characteristic of activity that accompanies any management decisions in conditions of uncertainty and is measured by the probability of not obtaining the expected results. That is, the results of risky decisions cannot be equated with a threat or danger, the action of which causes the deterioration of the system. Favorable conditions make it possible to obtain a result, to achieve the set goals, and only with a negative coincidence of circumstances, an undesirable result is possible.

As for the study of the content of the concepts of "threat" and "danger," the scientific literature allows us to come to a conclusion about the lack of clear definition of these concepts. "Danger" and "threat" are identified by some scientists; some define danger as a form of threat, and others define a threat as a stage or form of danger A number of researchers emphasize the consequences of threats and dangers, defining them as the action of factors, phenomena, events that cause an undesirable state of the enterprise, losses, damage, and so on.

So, from the author's point of view, a threat is a cause, phenomenon, event, or condition that can prevent the achievement of management goals and objectives. The situation becomes dangerous when the threat takes real forms of manifestation. That is, danger is a concrete form of manifestation of a threat, the destructive influence of which is fully realized and indisputable.

Returning to the results of the research on the content of risk, it is necessary to note its role in determining threats and dangers in the field of public administration.

Comparing the above concepts of risk and threat, it is possible to note that they are closely related. If a threat is a probable possibility of an obstacle to the goals and tasks of public administration, then the risk measures, evaluates this threat, and gives an idea about the mathematical expression of the probability of the occurrence of such obstacles.

Given that risk accompanies any management activity, threats are also an invariable attribute of the entire management system in the state. The degree of threat and its transition into a form of danger for management, in our opinion, is characterized by the level of risk, which, in turn, is determined by the probability of occurrence and the impact on management activities. That is, at an acceptable level of risk, the negative influence of external and internal environmental factors is considered a threat to activity, and when a critical level of risk is reached, the threat takes the form of danger.

Note that there is a direct dependence of the level of risk on the probability of negative events and the impact on management activities, but for each type of risk (each negative change in factors of the external and internal environment), the trajectory of the level of risk will have an individual form, which must be determined on the basis of mathematical modeling.

Thus, risks and threats in the field of public administration are of a constant nature with different levels, and the danger arises only when this level of critical importance is reached.

The variety of threats in the activity of a modern enterprise determines the need to systematize them according to certain key features, which will make it possible to timely diagnose them, assess them, and apply measures to overcome them or adapt to their impact.

The study of the essence of threats, their differences, and common features with risks and dangers, the proposed classification of threats, and their detailed description makes it possible to form a basis for diagnosing negative impacts from the stage of occurrence to overcoming. The system of tracking and assessing the characteristic signs of threats, their dynamics, and composition, is a necessary task of the management: a continuous process of identifying early signs of threats allows timely response to negative trends, preventing the spread of their impact on the companys activities or, in the absence of the possibility of prevention, "mitigating" their negative effect.

As for the content of the concepts "risk," "threat," and "danger," certain differences in their understanding can be distinguished: risk is an objective characteristic of the enterprise's activity, which accompanies any management decisions in conditions of uncertainty and is measured by the probability of not obtaining the expected results; threat causes, phenomena, events, and conditions that can prevent the achievement of goals and objectives of economic entities; danger is a specific form of manifestation of a threat, the destructive effect of which on the activity of the enterprise is fully realized and indisputable.

Threats and risks are indispensable attributes of management activity. The danger arises when the level of risk of critical importance of the entire system of public management and regulation in the state is reached.

Our conducted research allows us to state that risk in the field of public administration is a rather broad concept, a specific subject of scientific research that has its own status. In order to give a reasonable definition, it is necessary to find out what, in fact, is a risk in the field of public relations in public administration.

We believe that the objectivity of risk in the field of public administration is that it exists as a result of objectivity, inherent in management categories of conflict, uncertainty, vagueness, lack of comprehensive information at the time of assessment, and making certain management decisions. In turn, it is possible to distinguish and take into account several types of uncertainty and the socio-economic risk caused by it: (1) uncertainty of goals and criteria and, therefore, the need to take into account multicriteria in the processes of evaluation, management, decision-making; (2) lack of data, in particular quantitative data, which are necessary at the time of decision-making to calculate quantitative estimates of relevant indicators; (3) lack of time to make scientifically based decisions; (4) uncertainty of the actions of counterparties and competitors; (5) the ambiguity of estimates of forecasts of the development of the controlled object and the socio-economic environment.

The subjectivity of risk is caused by the fact that real people with their own experience, psychology, interests, tastes, manner of behavior, propensity or aversion to risk, etc., are involved in public activities. Thus, the objective-subjective nature of risk is determined by the fact that it is generated by processes of both subjective and objective nature, that is, those whose existence does not depend on the will and consciousness of people. As a subject of scientific research, the category "risk" should have its own object, subject, and sources of origin. The analysis of scientific literature devoted to the theoretical aspects of risk research made it possible to generalize the definition of such concepts. The object of risk can be understood as a public system and administration, and it is impossible to evaluate the effectiveness and conditions of its functioning in the future in exhaustive completeness and with necessary accuracy. A risk subject can be understood as a person or a team that is interested in the results of risk object management and has the appropriate competence to manage and make relevant decisions regarding the risk object. Under the sources of the origin of risk are meant factors, processes, and phenomena that cause uncertainty and conflict.

The study of the practical activities of subjects of public management activity gives reasons to claim that the main sources of the origin of risks include the following: (1) spontaneity of natural processes and phenomena, natural disasters; (2) randomness; (3) clash of conflicting interests; (4) uncertainty; (5) incompleteness, insufficient information about the object, process, phenomenon to which the management decision relates; (6) limitation, insufficiency of material, financial, labor, and other resources needed for making and implementing management decisions; (7) the impossibility of unambiguous knowledge of the object at the existing levels and methods of scientific knowledge; (8) inevitable differences in mechanisms and tools, intentions, assessments, stereotypes, behavior, and so on [7].

It is advisable to consider the risk in both historical and managerial aspects. As a historical category, risk is a possible danger perceived by a person, as risk is historically connected with the entire course of social development. As society develops, namely, with the emergence of commodity-money relations, risk becomes a category of economics and management. In the event of a risk, subjects of economic activity can receive three types of economic results: negative (loss), zero (neutral), and positive (profit). It should be noted that risk, like any economic category, manifests itself through functions. In particular, the economic literature distinguishes such risk functions as innovative, regulatory (constructive and destructive), compensatory (stabilizing), precautionary, control, and so on.

Probably, that is why, there is no single system of classification of management risks in the literature on public administration, but on the contrary, there are many approaches, which, for the most part, are determined by the purpose and tasks of the activity.

Having analyzed such a structural characteristic as the nature of accounting, we can claim that risks are divided into external and internal. External risks are understood to be risks that are not directly related to the public activities of the governing bodies of an individual country and the country as a whole.

In turn, in the structure of external risks, the following subtypes can be distinguished, namely, risks of the management system in the country – risks associated with the internationalization of the economies of countries, which can lead to the instability of state power, the state system and legislation, ineffective public policy, sharp polarization of the interests of various social groups of the country, etc.; force

DOI: http://dx.doi.org/10.5772/intechopen.1004455

majeure – risks associated with war, floods, earthquakes, storms and other natural disasters, revolutions, coups, strikes, etc., which interfere with the effectiveness of public administration; transfer – risks associated with the impossibility of transferring funds to the creditor's country due to currency restrictions in the borrower's country; tax – risks associated with a possible reduction in budget revenues as a result of a change in tax policy or the size of tax rates in individual countries.

Internal risks are understood as the risks that arise as a result of the activities of the national governing bodies of a certain country and the country as a whole. Analyzing management activity, it is possible to distinguish subtypes of internal risks, namely, resource risks, caused by the lack of a margin of safety in relation to resources in the event of a change in the situation, which leads to a deficit in the state's economic system; organizational – risks caused by deficiencies in work organization, which include errors in planning and design, deficiencies in work coordination, weak regulation, incorrect supply strategy, errors in the selection and placement of personnel, disorganized marketing activities, and unstable financial position; and social – risks caused by the assessment of loyalty risks, that is, the attitude to work, the quality of the workforce, etc. Also, the most important group, in our opinion, are the risks of public management, which are grouped by such a classification feature as the nature of the consequences of using mechanisms and tools in the field of management.

We note that the risk of public management is a category that characterizes relations between management subjects regarding their perception of objectively existing uncertainties and conflicts in management, decision-making, and evaluation, which are complicated by possible threats and unused opportunities. As for the classification of risks in public activity, according to the results of the research, in accordance with the practical and methodological tasks of determining their types, it is possible to predict the existence of a classification of risks based on the following characteristics: (1) time of occurrence; (2) nature of accounting; and (3) the nature of the consequences.

The classification of risk management mechanisms in public activity involves the allocation of two groups of such groups: (1) traditional, which in turn, is divided into contractual and classical-calculated ones and (2) unconventional, divided into regulatory (financial and economic) and institutional mechanisms. Therefore, public administration encourages the emancipation of the creative activity of the individual and synergy of the activity of the community and society. The development of the conceptual and categorical apparatus and conceptual ideas of public administration is the basis for the identification of risk patterns, which is a component of the modern theory of public administration.

6. Conclusions and prospects for further research

The analysis and research carried out in the article allow us to draw the following conclusions.

Under the conditions of martial law in Ukraine and updating the processes of public management of measures of emergency situations, risks and threats are carried out taking into account a clear algorithm and comprehensively from a legal point of view. Mostly, such management and regulation are ensured and concentrated in the fields of administrative, environmental, and labor law.

We consider it a task of the scientific field of public administration, and it is necessary to carry out a thorough analysis and monitoring of modern features of the legal evolution of domestic legislation in relation to the public sphere in the conditions of martial law.

It should be noted that, in the process of solving the problem of maintaining and ensuring the safety of social relations and citizens from an organizational and legal point of view, the leading countries of the world gradually improved their own legislation regarding emergency (military) situations – from the development and adoption of legislation regulating actions in the conditions of the occurrence and development of specific cases of danger before the adoption of separate laws regulating general conceptual actions during the war.

In recent years, many events and meetings have been held in Ukraine aimed at improving the ideology (concept) of public administration in emergency situations, but it was quite difficult to foresee an open war on the territory of Ukraine with the Russian Federation. Such measures were conditioned by the fact that the issues of safety and protection of the population, as well as economic and cultural objects from the occurrence and development of emergency situations, risks, and threats, were almost not perceived as a component of state security, with the recognition of the priority of the protection of the population and territories from various threats, which may arise during the long-term development of the entire system in the state, in contrast to solving the problems of defense, which is focused specifically on wartime. Thus, at the current military stage of the development of Ukraine's economy, a search is being made for a rational management model for providing protection, taking into account the effective and functional principle of response on the part of the state in the conditions of the occurrence and development of emergency situations, instead of using the usual industry principle, which was previously used in the public administration system.

It should be noted that the modern formation of the public administration system in the conditions of martial law in Ukraine is accompanied by the presence of a significant number of conflicts of legislation focused on the regulation of emergency situations, as well as the actual functioning of the system of forecasting, prevention, detection, and neutralization of emergency situations at the state level, particularly in the direction of the use of anti-crisis mechanisms and tools of public administration in the state.

The public administration system of Ukraine is doomed to function for a long time under constant pressure caused by the aggressive policy of the Russian Federation, which increases and diversifies the levers of destructive influence on economic development in Ukraine. The long-term nature of the threats generated by the aggressive course of the Russian Federation makes it impossible to build good-neighborly bilateral relations.

We note that, in such a situation, it is necessary to provide favorable conditions for the development of the economy after martial law on a resource basis, to provide a favorable environment for investment and entrepreneurship, a stable basis for financial stability, effective tools for the state to implement its social functions under conditions of permanently high risks of external military threats, and unpredictable hybrid impacts of various types and levels. The relevant policy should focus on overcoming the key obstacles to the country's strategically oriented development caused by the consequences of the Russian military aggression.

Public policy priorities should be (1) development of the potential of the real sector of the economy, overcoming export and import dependence on the Russian Federation; (2) implementation of structural transformation of old industrial regions with modernization of traditional sectors of the economy to restore (within economic expediency) their competitiveness on world markets; (3) implementation of the European industrial policy toolkit, strengthening of Ukraine's subjectivity as a partner of the EU; (4) increasing directions of implementation of the integration of the Ukrainian transport system into European and international projects; (5) a sharp decrease in the level of dependence on trade with the Russian Federation in goods and services in strategic and strategic areas of the national economy; (6) closure of Russian markets; (7) improvement and renewal of the Association Agreement with the EU, particularly in the field of Ukraine's integration into the EU's single energy market; (8) the introduction of real tax benefits for the transformation of small mining towns, provided for by the Law of Ukraine "On Amendments to the Tax Code of Ukraine and other legislative acts of Ukraine on ensuring the balance of budget revenues" dated November 30, 2021, no. 1914-IX; (9) reforming management mechanisms and tools to ensure the national interests of Ukraine and prevent economic and civil security risks; (10) formation of a system of information collection and processing and formation of a modern database on temporarily occupied territories; (11) accumulation of information about the actions of the Russian Federation regarding the undermining of the sovereignty of Ukraine in the economic and other spheres, development of compensation mechanisms for losses; and (12) institutional improvement of sanctions policy, particularly, its legislative support, elimination of identified shortcomings in this area with the aim of improving the interaction between state bodies and civil society, creating favorable conditions for effective response to existing and potential threats to the national security of Ukraine.

In the context of the analysis of post-war development priorities, intermunicipal and inter-regional cooperation can be considered as a powerful resource tool, which should solve the triple task of the national level: to promote the reintegration and consolidation of the country, to ensure the infrastructural integrity of the country, capitalization, mobilization, and optimization of the use of regional resources as the main source of economic reproduction of Ukraine.

Thus, the analysis of the current state, risks, and threats of the public administration system in relation to issues of improvement of anti-crisis mechanisms gives grounds for the conclusion that an imperfect system of legal regulation is currently operating in Ukraine, a sufficiently formal and categorical principle is dominant in accordance with the legislative norms of the pre-war state. While the regulatory system itself has a very complex structure and practically does not meet the military needs of the state. Therefore, the complexity of the object of legal regulation and the variety of problems create certain risks and threats, complications in the adjustment of legal norms in wartime. As a result, there are disagreements in different laws on the same issue, which complicates the possibility of their direct application in the field of resource provision during martial law and the period of further reproduction. A lot is being done today in this direction, and the experts of the Institute of Legislation are actively participating and providing proposals for amendments to the legislation in the conditions of martial law.

In our opinion and the opinion of experts-scientists, the rule-making work in this area still has reserves and is still proceeding by solving individual issues, without the necessary degree of generalization, which leads to contradictory trends in the process of legal regulation of public management and administration issues in the conditions of martial law in Ukraine. The sustainability of relations in this area is made impossible by the numerous changes that were constantly made to the current laws and other normative legal acts and by the lack of adaptation to the martial law. This leads to the need for a thorough study (monitoring) of the problems of legal regulation of public management and administration issues, in particular, issues related to the organization of the implementation of their functions by the relevant bodies, particularly ensuring the process of formation and use of all anti-crisis potential in the conditions of martial law and the subsequent recovery of the economy of Ukraine after the Victory over Russia.

Additional information

ORCID 0000-0002-4545-612X.

Author details

Valerii Ye. Vorotin
State Ecological Academy of the Ministry of Environment, Department of Public Administration, Kyiv, Ukraine

*Address all correspondence to: vevorotin@gmail.com

References

[1] List of All documents/Legislation of Ukraine [Internet]. Official website of the Parliament of Ukraine. 2022. Available from: https://zakon.rada.gov.ua/laws/main?find=2&user=a&org=1&typ=1%7C2&dat=20220224&datl= [Accessed: February 2, 2024]

[2] The Ukrainians will push the Russians out of the captured territories. The famous American philosopher Francis Fukuyama names the reasons why Ukraine will win [Internet] Available from: https://nv.ua/ukr/opinion/recommends/ukrajina-peremozhe-frensis-fukuyama-pro-te-chim-zakinchitsya-viyna-rosiji-ostanni-novini-50236988.html

[3] Pavlyuk O. The World Bank Has Preliminarily Assessed Ukraine's Losses as a Result of the War: It Is 40% of GDP [Internet]. Suspilne Ukraine; 2022. Available from: https://suspilne.media/231337-svitovij-bank-poperedno-ociniv-zbitki-ukraini-vnaslidok-vijni-ce-40-vvp/ [Accessed: February 2, 2024]

[4] Gaetani F. The Structure, Role and Mandate of Civil Protection in Disaster Risk Reduction for South Eastern Europe. South Eastern Europe Disaster Risk Mitigation and Adaptation Programme. Report International Centre on Environmental Monitoring Research Foundation (CIMA), Gaetani F, Parodi A, Siccardi F, Miozzo D, Trasforini E. [Electronic source]. Available from: http://www.unisdr.org/preventionweb/files/9346_Europe.pdf

[5] Vorotin Valerii, Sirinova Aliya. The Risk Management as an Innovation Approach in Govermental Management of Customs Procedures in GUAM Countries. Kyiv: dp "Vydavnychyi dim "Personal"; 2019;**2**(17):46-57

[6] Zhalilo Ya A, Bazylyuk Ya B, Sobkevich OV. The Strategic Price of Russian Aggression for the Economy of Ukraine [Internet]. National Institute of Strategic Studies; 2022. Available from: https://niss.gov.ua/sites/default/files/2022-04/analytrep_05_2022.pdf [Accessed: February 2, 2024]

[7] Vorotina N, Koval O, Vorotin V, Prodanyk V, Shynkarov A. Legal foundations and features of public administration in the budgetary sphere in Ukraine and abroad. Rivista di scienze giuridiche a cura della facoltà di giurisprudenza dell'università cattolica di Milano/vp vita e pensiero indice. 2023;**2**:247-262. (Scopus). DOI: 10.26350/18277942_000122

Chapter 10

Street Trading Uncovered: A Comprehensive Evaluation of Occupational Risk and Management Strategies in the South African Urban Marketplace

Maasago Mercy Sepadi and Lebogang Cleopatra Phama

Abstract

Street vending is a prevalent form of informal employment in South Africa, with both positive and negative impacts on workers' well-being and safety. Negative effects include psychological, respiratory, and musculoskeletal disorders. This chapter addresses health and safety concerns faced by street vendors in the informal trading sector using a systematic and scientific approach. Street vendors face unique challenges, such as navigating busy streets and dealing with extreme weather conditions. With changing climate and environmental factors, it is crucial to assess and manage these risks effectively. By conducting thorough risk assessments, potential hazards can be identified and strategies can be developed to mitigate them. Informal street trading is essential for urban economic growth and poverty alleviation. However, many businesses conducted by street vendors do not comply with workplace health and safety requirements, leading to high hazards and risks. Therefore, occupational safety and personal health risks faced by street vendors are crucial and require regulation of health and safety standards in the street trading industry.

Keywords: street trading, health, risk management, informal workers, exposure, assessment, street vendors, informal trading

1. Introduction

Street vendors play a crucial role in many communities, providing convenience and access to a variety of goods and services. As with any occupation, there are certain health and safety risks associated with this line of work. To establish a clear understanding, it is essential to define street trading activities. Selling goods or services in public spaces such as parks, sidewalks, or designated stalls is known as street trading. It is predominantly conducted by individuals or small-scale businesses, often without a fixed establishment. Street traders offer a wide range of products, including food, clothing, crafts, and various other merchandise. These activities involve direct interaction with customers in outdoor settings, making it a distinct form of commerce that requires specific consideration for health and safety regulations.

Street vendors often work informally and without fixed premises, relying on their mobile setups or carts to carry their products [1]. A study conducted in India found that street vending contributes to social, economic, and environmental sustainability by providing affordable food and creating job opportunities [2]. The informal trading sector benefits those who rely on it for economic benefits due to the scarcity of job opportunities, it provides competency, and it is also necessary for livelihood support [3].

Informal street vendors are prevalent in many urban areas, particularly in developing countries, due to increasing urbanization and industrialization. The exact prevalence varies depending on the specific geographical location and socio-economic conditions. The informal nature of their business makes it difficult to obtain accurate data on the total number of street vendors but estimates suggest that millions of individuals worldwide rely on street vending as their primary source of income. Furthermore, informal trading happens due to various reasons, for example economic benefits, high unemployment rates, using cheap and easily accessible sources of energy to prepare food, unregulated businesses, lack of resources to formalize such businesses. They operate outside of formal establishments such as shops or markets, setting up temporary stalls or selling directly from carts or their hands. These vendors often lack access to proper working spaces, basic amenities, and appropriate infrastructure. Their business operations are typically unregulated, making it challenging for authorities to monitor and control their activities. Currently, informal food traders' premises are not regulated in South Africa in terms of R638 (Regulations Governing General Hygiene Requirements for Food Premises, the Transport of Food and Related Matters), especially in defining the requirements of the structure. This results in an influx of informal food traders in areas such as metropolitan cities because of informal food preparation. All activities that take place within the occupational environment should be regulated in order to put control measures in place to preserve the environment and prevent any negative health implications for the people at large [4].

The health exposures and outcomes of informal street vendors vary based on the type of vendor and the location of their stalls [5, 6]. Activities that are undertaken in the informal trading sector include food preparation using charcoal as a source of energy, and selling of toys, cosmetics, clothes, music, books, furniture, etc. These activities usually take place in areas that are not zoned for these activities. Some activities that are undertaken in the informal food sector contribute to the emissions and transmission of air pollutants in the atmosphere [7]. Different types of air pollutants can deteriorate the air quality standards in urban areas depending on the concentration and the potential toxicity of these pollutants and may also affect the health of the communities at large. There is an increase in urbanization and industrialization in metropolitan cities such as Johannesburg as people seek better livelihoods. Many of these people conduct businesses that are not regulated which results in environmental health challenges such as air pollution. One such business is informal food trading, which is conducted on pavements, open veld, sidewalks, unauthorized spaces, and taxi ranks. Food is usually prepared using the cheapest and easiest methods such as broiling and braaing using charcoal as a source of energy. When braaing and broiling take place, three combustion phases are involved namely, igniting, flaming, and coking. During these combustion phases, smoke is released into the atmosphere resulting in the potential inhalation of different volatile organic compounds (VOCs) such as benzene, toluene, ethylbenzene, and xylenes (also known as BTEX) and these are harmful to the environment and the people. Research suggests that the emission of smoke resulting from charcoal burning comprises pollutants such as VOCs, particulate matter, carbon monoxide, nitrogen dioxide, and toxic organic compounds such as benzene, 1, 3 butadiene, and many other PAH [8].

Ensuring the safety and well-being of street vendors is crucial in today's rapidly changing world. With occupational safety, environmental health, personal health, and climate change becoming increasingly important factors to consider, risk assessment, and management play a vital role in safeguarding the livelihoods of these hardworking individuals. The purpose of this document is to assess the risks faced by street vendors in their daily work activities. Street vending is a common form of informal employment in many countries, providing income for a significant portion of the population. However, it also presents various risks and hazards that can affect the health and safety of vendors. By conducting a thorough risk assessment, we aim to identify and address these risks to ensure a safe working environment for street vendors. Street vendors play a significant role in urban economies, providing goods and services to large sections of the population. However, their work is not without challenges. This content focuses on the environmental risks faced by street vendors and their implications. By understanding the various risks, we can explore potential solutions and strategies to mitigate these risks and improve the working conditions for street vendors.

This chapter aims to highlight the common health and safety emergencies that street vendors may face and provide information on how to prevent and respond to them. By understanding these risks and taking appropriate measures, street vendors can ensure the well-being of themselves, their customers, and the community they serve. To ensure the safety of street vendors, it is essential to conduct regular risk assessments that consider the specific hazards they face. This includes identifying potential sources of harm such as heavy traffic, hazardous materials, or unsanitary food handling practices. By prioritizing occupational safety and environmental health for street vendors, we not only protect their well-being but also contribute towards creating a healthier and more sustainable urban environment for all.

2. Methodology

The scope of this risk assessment encompasses all aspects of street vending activities, including the physical, environmental, and occupational health and safety and personal health-related risks faced by vendors. It considers a wide range of factors such as the nature of the vending location, the materials and equipment used, the handling of foodstuffs or materials, and the interactions with customers and the public.

The risk assessment and management will be conducted using a systematic and scientific methodology. It involves literature from primary studies in which their data collection was conducted through site inspections, interviews with vendors, and consultation with relevant stakeholders. The researchers also investigated publications that examined available data on reported crimes where a street vendor was the victim.

3. Assessing potential risks and hazards in the street trading industry

The literature gives evidence that informal street vendors are exposed to harmful physical, chemical, microbiological, and ergonomic stressors. The purpose of this risk assessment is to systematically evaluate the potential risks faced by street vendors in order to develop effective mitigation strategies. By understanding the specific hazards and vulnerabilities associated with street vending, we can implement measures to prevent accidents, injuries, and illnesses. The assessment will provide valuable insights into the nature and severity of the risks involved, allowing for informed decision-making and targeted interventions to protect the health and safety of street vendors.

3.1 Physical hazards analysis

The physical hazards of vendors' work include Accidents and Injuries, Fire Hazards, Unsafe Equipment, and Structures. However, more needs to be done to address the structural inequities that hinder street vendors from making a living and a life [9].

3.1.1 Identifying vulnerabilities in street vending locations

Unfortunately, street vendors are also at risk of accidents and injuries. Street vendors are often exposed to violence and crime, especially against women street vendors, who in particular, are at a higher risk of violence and harassment on a daily basis and from a variety of sources. Los Angeles saw a nearly 37% increase in recorded crimes against street vendors between 2010 and 2019, from 38 to 166 [10]. Robberies accounted for over 45% of all offences against street vendors from 2010 to 2019, according to the LAPD. During that period, approximately 28% of crimes involved assault [10]. A corn vendor was beaten, robbed, and shot to death on Milwaukee's south side sidewalk, near 11th Street and Grant Street, according to Fox6 [11]. In August 2023, a cash-in-transit vehicle plowed into a group of street vendors in Durban, South Africa, leaving eight people injured [12].

In October 2022, gunmen opened fire on street vendors selling chicken feet in Johannesburg, South Africa, killing seven people and injuring four others [13–15]. Among the risk concerns were open fires and non-enclosed vendor stalls. A small section of the building on AB Xuma (Commercial) Street in Durban Central that is rented out to street vendors was damaged by the fire that erupted on February 9, according to a report by Timeslive [16]. In June 2022, a fire in Johannesburg, South Africa's Yeoville Market destroyed twenty-five stalls owned by some of the 208 street vendors there [13, 15].

3.2 Overview of environmental risks faced by street vendors

Street vendors are exposed to a range of environmental risks that can have detrimental effects on their health, safety, and livelihoods. This section provides an overview of these risks, including health hazards such as air pollution exposure and contamination of food and water. It also examines safety risks, including traffic hazards, risk of accidents and injuries, exposure to violence and crime, as well as fire hazards. A research article published in 2023 analysed the impact of street vending on urban ecology sustainability [2]. The study created an epistemic framework to comprehend the structure and function of urban street vending. According to the study, street vending is a necessary urban activity that emerges spontaneously rather than being "formally planned." Street vendors' activities are therefore considered "informal" and are not taken into account in traditional urban planning and design. Additionally, the study discovered that street selling affects the urban ecosystem in both positive and negative ways [2]. Additionally, considering the environmental impact of their activities is crucial in mitigating risks associated with climate change. Understanding these environmental risks is essential to develop effective interventions and support mechanisms for street vendors.

3.2.1 Air pollution health risks

Numerous threats to occupational health and safety exist for street vendors. According to a study by the International Institute for Environment and Development (IIED) [17], South African street food sellers face dangers at work that have an impact on their health and general well-being. However, street vendors are constantly exposed to outdoor air pollution from a variety of sources, such as traffic, unpaved roads, building sites, manufacturing facilities, inadequate ventilation, open fires, and biomass fuel from petrol [5, 6]. Street vendors are at risk of traffic hazards such as heavy traffic movement from different vehicles that emit pollutants such as VOCs, noise exposure, high-speed cars, and distractions. Vehicle exhausts contribute to high air pollutants especially in urban areas due to heavy and congested traffic [18]. Furthermore, the study was conducted during one season, and according to the literature, emissions are distinctively different depending on seasonal changes. Volatile organic compounds are harmful to human health even at low concentrations, and organs such as the liver, kidney, central nervous system, and respiratory system are affected by exposure to VOCs [19]. A better understanding of VOC distribution characteristics is important for discussion on the behaviour and fate of VOCs to better determine the causes of occupational diseases that can occur in the informal food trading sector. Ingestion, inhalation, and skin contact are major routes of exposure to BTEX [20]. Volatile organic compounds are also known to affect the immune system and can cause nausea and headaches and can cause haze and photochemical smog in the environment. Moreover, according to Majumdar et al. [19] VOCs are associated with severe asthma symptoms in children. Benzene, toluene, ethylbenzene, and xylene compounds are known to be harmful to human health, and these compounds can cause acute and chronic health effects on human health.

3.2.2 Climate change resilient trading

One of the groups most susceptible to the effects of climate change is street sellers. According to a 2016 report by the International Institute for Environment and Development (IIED), street vendors and other informal labourers are subject to a variety of climate-related dangers at work, including intense heat waves, flooding, pollution, and drought, susceptibility to meteorological conditions. Street vendors face a serious threat from rising temperatures. According to an IIED survey of 420 informal labourers in Harare and Masvingo, heat waves have caused over half of them to become exhausted, work fewer hours, and earn less money [9]. Furthermore, one of the groups most susceptible to the effects of climate change is street vendors. Rising temperatures pose a significant threat to street vendors, causing reduced working hours, exhaustion, and lower incomes for over half of them. Key informal economy groups are crucial in mitigating climate change, with outdoor street traders likely to leave a smaller carbon footprint [5, 6, 21].

Street vendors are exposed to various air pollutants such as volatile organic compounds (VOCs) that are widely distributed in a lower urban atmosphere, influenced by different anthropogenic activities such as combustion, intense transportation, and industrial and commercial activities especially in urban areas [19]. According to Majumdar et al., [19], the behaviour and fate of chemicals in the ambient air are determined by the physio-chemical properties, how the chemical is emitted, and the environmental conditions. In addition, VOCs have the potential to produce tropospheric ozone and other photochemical oxidants due to their chemical reactivity; therefore, they are considered contributors to global warming [19]. BTEX compounds are released into the air as combustion by-products of fossil fuels and biomass and are further classified as hazardous air pollutants and are regulated by agencies such as the United States Environmental Protection Agency.

In the face of climate change and its impact on our environment, it is crucial to address the risk management strategies for street vendors. Understanding of release and emissions of air pollutants is important for the implementation of regulatory frameworks as well as safer design consideration and control initiatives. Risk management should be process-specific. Pollution abatement control measures should be explored. These hardworking individuals are often exposed to extreme heat and other environmental challenges that can have detrimental effects on their health and livelihoods. Climate change has led to an increase in heat extremes, posing a significant threat to the well-being of street vendors. It is essential to develop comprehensive risk management plans that not only protect their health but also ensure that their businesses remain resilient in the face of changing climatic conditions.

One way to achieve this is by implementing effective health information systems specifically tailored for street vendors. These systems can provide valuable data on environmental health risks, allowing vendors to make informed decisions about when and where they should operate. By having access to real-time information, they can better protect themselves from extreme heat and other environmental hazards. Furthermore, promoting climate resilience among street vendors involves providing them with the necessary tools and resources. This could include access to shaded areas or portable cooling systems during hot weather conditions. Additionally, educating them about the importance of staying hydrated and taking regular breaks can significantly contribute to their overall well-being.

3.3 Occupational health and safety risks

Health and safety play a crucial role in street trading activities. Ensuring the well-being of both street traders and the general public is of paramount importance [22]. Street trading environments present various potential risks and hazards that need to be mitigated. By promoting health and safety measures, we can minimize accidents, injuries, and illnesses. These measures include maintaining proper hygiene practices, understanding, and preventing fire hazards, following food safety standards, and implementing accident prevention strategies. Prioritizing health and safety not only protects those directly involved in street trading but also contributes to the overall well-being and reputation of the community.

3.3.1 Identifying occupational health and safety hazards

Workplace hazards and risks include physical, microbiological, chemical, ergonomic, and psychological, stressors. Workplace psychological stressors must be given priority since the public health problem of mental health disorders outweighs the resources allocated to its treatment. The occupational health hazards faced by vendors include exposure to harmful substances, musculoskeletal disorders, mental health challenges, and fatigue and stress management. Hazardous and biological substance challenges include poor storage of chemicals, usage of outdated chemicals/ substances, and selling of illegal substances, for example pest control. Ergonomic stressors can be attributed to inadequate structures, lack of training and education, insufficient health and risk assessments, and exposure to physical hazards. Ablution facilities are not in sanitary conditions, are inadequate, and do not meet the standards in facilities regulations, no hand wash basins, and inadequate water points for supplying water for hand washing and cleaning purposes. The Mental Health Effects may include Stress and Psychological Well-being, Social Isolation and Loneliness, Absenteeism, and poor working relationships among colleagues.

3.4 Economic and social risks

Street vendors often experience insecurity of livelihood, lack of legal protection and social security, and Weather-Related Emergencies.

3.4.1 Common health and safety emergencies faced by street vendors

The introduction of this work on potential health and safety emergencies in street trading activities aims to provide an overview of the topic. It will explore the various risks and challenges that street traders may encounter in their day-to-day operations. By understanding the potential hazards, traders can take appropriate measures to protect themselves and the public. This work will delve into the importance of health and safety in street trading, common emergencies that can arise, the need for emergency preparedness and response, and the regulatory framework and compliance requirements that govern street trading activities. Common health and Safety emergencies include Fire hazards and prevention, Foodborne illnesses, and hygiene practices, and Physical injuries and accident prevention.

Street vendors face several unique challenges when it comes to emergency preparedness. These challenges include limited access to resources and information, lack of training and knowledge on emergency response protocols, difficulty in relocating businesses during emergencies, and potential language barriers in communicating with authorities and customers. Additionally, street vendors often operate in densely populated areas with limited infrastructure support, increasing their vulnerability to natural disasters and other emergencies. Recognizing and addressing these challenges is crucial in developing effective emergency preparedness strategies for street vendors. Emergency preparedness for street vendors is a vital aspect of ensuring the safety and livelihoods of these individuals in times of crisis. With the potential for various emergencies to occur, street vendors must be equipped with the necessary knowledge and resources to protect themselves, their businesses, and their customers. This work aims to provide comprehensive guidance on emergency preparedness specifically tailored to the unique circumstances and challenges faced by street vendors.

4. Impact on public health

Informal street vendors play a crucial role in urban economies around the world. These vendors typically sell a wide range of goods and services on the streets, often without proper licenses or permits. While their presence contributes to the local economy, there are also significant health risks associated with their activities. This section aims to explore the various environmental health hazards faced by informal street vendors and the impact they can have on public health.

Many obstacles prevent street food vendors from improving the microbiological quality of their offerings [23]. Food contamination is commonly attributed to exposure to unhygienic conditions, a lack of amenities, such as readily available potable water, and inadequate hygiene procedures on the part of food handlers [23]. Foodborne illnesses, the rise of infectious diseases, respiratory issues, and an increase in chronic diseases are some further effects on public health. Furthermore, street sellers are always exposed to outdoor air pollution from a variety of sources, including automobile and traffic emissions, unpaved roads, building sites, industrial activities, limited or poorly ventilated stall spaces, and biomass fuel from open fires and gas stoves [5, 6].

A comprehensive evaluation of original publications published between January 2010 and July 2023 discovered that street food vendors in underdeveloped countries had a low level of knowledge and were unaware of proper food hygiene measures [23]. There is also a risk of foodborne illness; the review also found that *Enterobacteriaceae* such as *Escherichia coli* (61.9%), *Salmonella* (30.1%), and *Shigella* spp. (9.5%), as well as *Staphylococcus aureus* (30.1%) and *Listeria monocytogenes* (14.3%), were the most common pathogens found in ready-to-eat street foods [23]. According to the FAO and World Health Organization (WHO) [24], street food vendors' lack of understanding about the causes of foodborne infections is a serious public health risk factor. Many developing nations continue to face substantial challenges in successfully implementing street food vending policies that are aligned with public health goals ([23]; FAO and WHO [24]). Vendors are susceptible to gastrointestinal disorders such as salmonellosis and respiratory infections such as influenza and COVID-19 due to a lack of access to water, waste disposal facilities, and basic hygiene understanding and practices. Air pollution raised the danger of respiratory and urinary infections, as well as affecting female street vendors' reproductive health.

Other public health challenges within the informal street trading/street vendors include:

- *Waste management:* poor waste management practices, no refuse bins, waste collection not done regularly, no designated waste areas pending collection, inadequate bins within, waste not segregated according to its type (garden waste, kitchen waste, healthcare waste, electronic waste, rubble, etc.)
- *Environmental degradation:* environmental decay, degradation, effluent, run-offs, unhygienic conditions, stagnant water breeding place for rodents, no proper structure that can prevent environmental pollution, etc.
- *Air pollution impact:* Emissions from energy sources that are used, for example charcoal, wood, papers, and burning of tyres to keep warm in winter. Emissions from combustion phases deteriorate air quality standards which have the potential to cause harm when exposure happens.
- *Food safety:* counterfeit foods, poor cold chain, no washing of hands, cooked foods mixed with raw foods, unregulated supply of meat products, food sampling is not conducted from street vendors, food products not labelled, selling of expired foods, etc. [25]
- *Pest control:* breeding places for rodents, unsanitary conditions, an environment full of pests and untreated.

- *Diseases and illnesses:* unhygienic practices lead to the multiplication of bacteria, microorganisms, viruses, etc., vector-borne diseases, waterborne diseases due to drinking contaminated water and ingesting contaminated and unsound foods.

5. Risk management strategies

Street vending is a common source of food for many people in developing countries. By prioritizing risk assessment and management for street vendors, we demonstrate our commitment to creating a safer working environment for these individuals who contribute significantly to our communities. Together, we can empower them with the necessary tools and support needed to navigate the challenges they face while ensuring their sustainable livelihoods for years to come. Therefore, risk management strategies need to be implemented taking into consideration factors such as hazards or stressors, exposure pathway, type of stressors, the concentration of stressors, factors that exacerbate stressors concentration/survival, exposure assessment, risk characterization, appropriate control measure, legal considerations, other stakeholders, socio-economic factors, etc.

5.1 Occupational health and safety management

It is imperative that we work together towards creating a more resilient environment for these hardworking individuals who play a vital role in our communities. Several cities have implemented policies and programmes to improve access to sanitation facilities for street vendors. Here are some examples, such as Kumasi, Ghana the city that invested in pay-per-use public toilets to meet the demand from traders, transient workers, and migrants for proper sanitation services [26]. In Mumbai, India a programme was implemented called "Clean Mumbai Green Mumbai" that aims to improve access to sanitation facilities for street vendors [27].

Local governments can take several measures to improve access to sanitation facilities for street vendors. Here are some ways:

- *Developing health and safety regulations for street trading and implementing policies:* Local governments can develop and implement policies that promote the provision of clean water and sanitation facilities for street vendors. These policies should be aligned with public health policies and should be enforced by relevant authorities [28]. There is a need to establish occupational health and safety standards for informal work such as street trading.

- *Building partnerships:* Local governments can build partnerships with non-profit organizations, community-based organizations, and other stakeholders to improve access to sanitation facilities for street vendors. These partnerships can help to mobilize resources, share knowledge, and build capacity [29].

- *Decentralizing services:* Local governments can decentralize services to improve access to sanitation facilities for street vendors. This can involve the establishment of decentralized water and sanitation systems that are managed by local communities [29]

- *Utilizing technology:* Local governments can utilize technology to improve access to sanitation facilities for street vendors. For example, mobile applications can be developed to provide information on the location of public toilets and water points [29].

- *Promoting community participation:* Local governments can promote community participation in the planning and implementation of sanitation facilities for street vendors. This can involve the establishment of community-based committees that are responsible for the management of sanitation.

- *Allocate sites/areas and combat crime:* Areas should be allocated to ensure that informal trading takes place in designated areas or spaces and ensure that those designated areas are safe and secured.

Furthermore, effective risk management strategies should be implemented to mitigate these risks. This may involve providing appropriate protective equipment, improving access to clean water and sanitation facilities, or implementing measures to address climate change-related challenges. Occupational safety measures such as providing adequate training on handling equipment and implementing proper hygiene practices are essential for protecting street vendors from physical injuries and illnesses.

5.2 Environmental impact and climate change control

By prioritizing risk management strategies for street vendors in the context of climate change, we not only safeguard their health but also support their economic sustainability. It is imperative that we work together towards creating a more resilient environment for these hardworking individuals who play a vital role in our communities.

- *Using alternative energy sources:* Street vendors can reduce their reliance on fossil fuels by utilizing alternative energy sources such as solar power or biogas, which can help mitigate climate change impacts and lower greenhouse gas emissions.

- *Adopting sustainable practices:* Street vendors can promote sustainability by utilizing biodegradable packaging, minimizing food waste, and conserving water [30]

- *Increasing networks and advocating for vendors' rights in climate change:* Street vendors can build networks with other vendors and community organizations to share knowledge and resources. This can help to build resilience and adapt to the impacts of climate change. Street vendors can advocate for their rights and demand that local governments take action to address the impacts of climate change. This can involve participating in policy-making processes and building alliances with other stakeholders.

5.3 Emergency preparedness for street vendors

There is a need to develop an emergency preparedness plan, establish communication channels, establish emergency response procedures, train street vendors on emergency preparedness, implement and maintain preparedness measures, stock essential emergency supplies, establish collaboration with local authorities, and conduct regular emergency drills and exercises. In the event of an emergency, street vendors should have a disaster preparedness plan in place. A comprehensive disaster preparedness plan should include knowing vendors, assessing facilities' ability to withstand the storm, evaluating utilities, and assessing personnel [31]. In addition to building relationships with vendors, street vendors can also take self-help measures to prepare for emergencies. For example, they can form cooperatives to pool resources and share knowledge [32].

5.3.1 Importance of emergency preparedness

Emergency preparedness holds significant importance for street vendors as it can make a critical difference in effectively responding to and recovering from emergencies. By being prepared, street vendors can minimize the impact of disasters on their businesses, reduce potential harm to themselves and their customers, and ensure faster recovery. It also demonstrates their commitment to the safety and well-being of their communities. Understanding the importance of emergency preparedness is crucial for street vendors to safeguard their livelihoods and contribute to the overall resilience of the local economy.

5.3.2 Objectives of the emergency response strategy for street trading

The primary objectives of this work are to enhance the emergency preparedness of street vendors and empower them to effectively manage potential emergencies. It aims to provide a comprehensive understanding of potential emergencies faced by street vendors, identify vulnerabilities in their vending locations, and assess risks and hazards unique to their circumstances. Additionally, this work seeks to guide the development of emergency preparedness plans, including establishing communication channels, creating emergency response procedures, and training street vendors on emergency preparedness. By achieving these objectives, we strive to ensure the safety and resilience of street vendors and their communities.

5.4 Healthcare for street vendors

Street vendors should be provided with access to social security benefits such as healthcare, pensions, and unemployment insurance [33]. Furthermore, addressing personal health concerns is equally important. Offering access to healthcare services and promoting healthy lifestyle choices can improve the overall well-being of street vendors while reducing the likelihood of long-term health issues. Local government can assist by conducting outreach programmes such as mobile healthcare services to promote and improve good healthcare practices among street vendors.

To protect themselves from these risks, street vendors can take several measures such as maintaining good hygiene practices, using protective equipment, staying hydrated, taking regular breaks, diversifying their products, using alternative energy sources, adopting sustainable practices, building networks, and advocating for their rights. However, street vendors often face challenges in accessing clean water and toilets, which can have a significant impact on their health and earnings. Local governments can take several measures to improve access to sanitation facilities for street vendors, such as developing and implementing policies that promote the provision of clean water and sanitation facilities, building partnerships with non-profit organizations and other stakeholders, decentralizing services, utilizing technology, and promoting community participation in the planning and implementation of sanitation facilities. All government spheres need to fast track the formalization of the informal sector to ensure that laws and rules will be implemented to better govern all activities that are conducted in the informal sector.

5.5 Regulatory framework and compliance

Street vendors are recognized as business individuals under the Businesses Act, No 71 of 1991. South Africa's Street vendors lack legal protections and formal labour law rights, resulting in limited legal protections for them [34]. The current regulatory measures in street trading do not cover the occupational health and safety of this industry. The Businesses Act No. 71 of 1991 established the National Framework for Street Vendors, officially recognizing street vendors as business owners. However, the non-labour regulation under which they operate is often inappropriate to their situation or improperly applied, often leading to harassment [33, 34]. More legislative frameworks that should be complied with by street trading include the Constitution of the Republic of South Africa, the National Health Act, the Environmental Management Act, the National Waste Management Act, Local By-Laws, Foodstuffs, Cosmetics and Disinfectant Act, Occupational Health and Safety Act, Hazardous Chemical Substances Act, Facilities Regulations and SANS-Water 241.

The local government in South Africa plays a crucial role in regulating street vending. There is a need for improvement in compliance by street vendors with local regulations and permits for street trading. There is also a need for government to ensure health and safety inspections and enforcement, and compliance with food safety standards. To improve the regulatory framework for street vending in South Africa, it is recommended to:

- *Recognize street vendors as workers:* Street vendors should be recognized as workers and be granted the same rights and protections as other workers in South Africa [33].

- *Provide legal protections*: Street vendors should be granted legal protections under the labour law system in South Africa. This would ensure that they are not exploited and are treated fairly by their *employers* [33].

- *Provide training and support:* Street vendors should be provided with training and support measures to help them run their businesses effectively and efficiently. This can be done through formal education and awareness campaigns. This would include training in business management, marketing, and financial management.

- *Regular law enforcement:* Regulation and enforcement should not be done per department but be joint work with all law enforcement stakeholders to avoid confusion of requirements for compliance.

- *Create partnerships:* Street vending in South Africa is a vital part of the informal economy, promoting local economic development. Partnerships between street vendors, local governments, and stakeholders can provide resources such as microloans, training programmes, and marketing support while advocating for vendors' rights and establishing legal frameworks.

Local economic development is a process where municipal authorities, community-based organizations, and local communities stimulate economic activity to create employment. It involves building on local resources, including human, capital, and institutional resources. South Africa's constitution mandates municipalities to drive local social and economic development, as outlined in the White Paper on Local Government and the Municipal Systems Act. A mutual understanding between street vendors and local authorities can enhance local economic development. This successfully directs municipalities to form partnerships that can assist them enhance the livelihoods of local residents. It also authorizes towns to use legal and other tools to foster the growth of both official and informal businesses [35].

6. Research limitations

Insufficient research conducted in the informal street trading and street vendor sector focusing on occupational health and safety in South Africa resulted in low science-based evidence, which in turn can help improve working conditions. Exposure to different stressors found in street vendors will vary according to meteorological conditions; there it is important to consider weather differences when conducting health and safety assessments. Furthermore, there is no police crime and health and safety incident data related directly to street trading or available literature documenting South African street trading industry crime, which makes it difficult to identify and compare incidents occurring in this trade. Results of the research indicate the need to assess control measures in relation to different street trading activities conducted. There are no direct legal frameworks or regulations to ensure compliance with occupational health and safety requirements in the street trading industry, and this gap limited the outcome of this research. There is a need for the establishment of street vendors' health and safety compliance standards which will assist in preventing workplace-related incidents and accidents.

7. What gaps does this research fill?

This research provides a thorough risk evaluation of the occupational health and safety challenges in street trading. Furthermore, it contributes new management strategies that could be adopted for the management of street trading in cities.

8. Conclusions

Street vendors are crucial contributors to the economy, particularly in urban areas, as they provide employment and income, meet demand for affordable goods and services, and foster entrepreneurship. This helps improve the livelihoods of local communities. Street vendors often operate in challenging conditions, facing exposure to extreme weather conditions, pollution, and inadequate sanitation facilities. These factors can have a detrimental impact on their health and overall well-being. These stressors have a great impact on the health of the informal street traders, and they affect the quality of the environment. Ensuring clean and safe food handling practices reduces the likelihood of foodborne illnesses and the harmful effects of contamination. By prioritizing risk management strategies for street vendors in the context of climate change, we not only safeguard their health but also support their economic sustainability. Addressing physical hazards such as accidents, fires, and unsafe equipment promotes a secure working environment. Additionally, recognizing and managing occupational health risks such as exposure to harmful substances, musculoskeletal disorders, mental health challenges, and fatigue can safeguard the overall well-being of street vendors. By acknowledging the importance of health and safety, street vendors can enhance the quality of their work while fostering trust and confidence among their customers and the community.

Health and safety should be of paramount importance for street vendors due to the potential risks and hazards involved in their daily operations. By prioritizing health and safety, street vendors can protect themselves, their customers, and the public. Interventions in controlling and mitigating stressors in the informal sector need to be process-specific and need to be grouped and categorized according to their effects and possible health outcomes. Scientific evidence is needed to ensure that appropriate and effective interventions are instituted. Health and risk assessments need to be conducted by various stakeholders focusing on different aspects that contribute to the exposure of the street vendors and implementing interventions that will control stressors.

Authors' contributions

MMS conceptualized and established the methodology of the study and wrote the original draft. MMS and LCM collected data and refined the manuscript. Review and editing of the manuscript were done by MMS and LCM. All the authors have read and agreed to the final manuscript.

Conflict of interest

The authors declare no conflict of interest.

Author details

Maasago Mercy Sepadi[1*] and Lebogang Cleopatra Phama[2]

1 Faculty of Sciences, Department of Environmental Health, Pretoria Campus, Tshwane University of Technology, Pretoria, South Africa

2 City of Johannesburg Metropolitan Municipality Region G, Environmental Health Department, Ennerdale Civic Centre, Ennerdale, South Africa

*Address all correspondence to: sepadimm@tut.ac.za

References

[1] Lemessa SD, Watabaji MD, Yismaw MA, et al. Evening street vending and the tragic living conditions of vendors: The case of eastern Ethiopia region. Cities. 2021;**108**:102947. DOI: 10.1016/j.cities.2020.102947

[2] Isha P, Pratap R. Impact of street vendors on urban ecology sustainability. Environment and Ecology Research. 2023;**11**(6):891-903. DOI: 10.13189/eer.2023.110601

[3] Young G, Crush J. Governing the informal food sectors in cities of the global south (Discussion papers no. 30)', (30). 2019. Available from: http://hungrycities.net [Accessed: January 28, 2023]

[4] Buluş E, Sakarya Buluş G, Yakuphanoglu F. Production of polylactic acid-activated charcoal nanofiber membranes for COVID-19 pandemic by electrospinning technique and determination of filtration efficiency. Journal of Materials and Electronic Devices. 2020;**4**:21-26

[5] Sepadi MM, Nkosi V. Environmental and occupational health exposures and outcomes of informal street food vendors in South Africa: A quasi-systematic review. International Journal of Environmental Research and Public Health. 2022a;**19**(3):1348

[6] Sepadi MM, Nkosi V. Street Food Keeps Johannesburg Going - but Working Conditions of Vendors Are Unhealthy. Braamfontein, South Africa: The Conversation Africa, Inc.; 2022b. Available from: https://theconversation.com/street-food-keeps-johannesburg-going-but-working-conditions-of-vendors-are-unhealthy-194089 [Accessed: January 01, 2024]

[7] Mohlakoana N, de Groot J, Knox A, Bressers H. Determinants of energy use in the informal food sector. Development Southern Africa. 2019;**36**(4):476-490. DOI: 10.1080/0376835X.2018.1526059

[8] Juntarawijit C, Juntarawijit Y. Cooking with biomass fuel and cardiovascular disease: A cross-sectional study among rural villagers in Phitsanulok, Thailand. F1000Res. 2020;**9**:307. DOI: 10.12688/f1000research.23457

[9] International Institute for Environment and Development (IIED). Green economy must work for street vendors as well as wall street. 2016. Available from: https://www.iied.org/green-economy-must-work-for-street-vendors-well-wall-street [Accessed: January 01, 2024]

[10] Crosstown. Street vendors risk a lot to make a living. 2020. Available from: https://xtown.la/2020/10/08/street-vendors-crime-los-angeles/

[11] Fox6. An honest man working: Popular corn vendor beaten, robbed, shot to death on Milwaukee's south side. 2020. Available from: https://www.fox6now.com/news/an-honest-man-working-popular-corn-vendor-beaten-robbed-shot-to-death-on-milwaukees-south-side [Accessed: January 01, 2024]

[12] East Coast Radio. Watch: Cash van ploughs into street vendors in Durban. Online. 2023. Available from: https://www.ecr.co.za/news/news/watch-cash-van-ploughs-street-vendors-durban/ [Accessed: January 01, 2024]

[13] News24. Seven dead as gunmen open fire on Joburg street vendors. 2022a. Available from: https://www.news24.com/news24/southafrica/news/

seven-dead-as-gunmen-open-fire-on-joburg-street-vendors-20221030 [Accessed: January 01, 2024]

[14] BusinessLive. Seven killed as gunmen open fire on Ennerdale street vendors. 2022. Available from: https://www.businesslive.co.za/bd/national/2022-10-30-seven-killed-as-gunmen-open-fire-on-ennerdale-street-vendors/ [Accessed: January 01, 2024]

[15] News24. City of Joburg refurbishes Yeoville Market after devastating fire amid probe by cops, SAHRC. 2022b. Available from: https://www.news24.com/news24/southafrica/news/city-of-joburg-refurbishes-yeoville-market-after-devastating-fire-amid-probe-by-cops-sahrc-20220920 [Accessed: January 01, 2024]

[16] TimesLive. Future up in smoke for Durban informal traders after building fire. 2022. Available from: https://www.timeslive.co.za/news/south-africa/2022-02-24-future-up-in-smoke-for-durban-informal-traders-after-building-fire/

[17] International Institute for Environment and Development (IIED). Street vendors and garbage pickers need protection from climate extremes. 2021. Available from: https://www.iied.org/street-vendors-garbage-pickers-need-protection-climate-extremes [Accessed: January 01, 2024]

[18] Xuan L et al. Source, temporal variation and health risk of volatile organic compounds (VOCs) from urban traffic in Harbin, China. Environmental Pollution. 2021;**270**(3):116074. DOI: 10.1016/j.envpol.2020.116074

[19] Majumdar D, Mukherjeea AK, Sen S. BTEX in ambient air of a metropolitan city. Journal of Environmental Protection. 2011;**02**(01):11-20. DOI: 10.4236/jep.2011.21002

[20] Dadban Shahamat Y, Khani MR, Ghahrchi M. Description of BTEX concentrations with wind roses in Tehran city: The study of monitoring, GIS-zoning maps and risk assessment. Human and Ecological Risk Assessment: An International Journal. 2021;**27**(9-10):2311-2327. DOI: 10.1080/10807039.2021.1999203

[21] ClimateHomeNews. Street vendors and garbage pickers need protection from climate extremes. 2021. Available from: https://www.climatechangenews.com/2021/09/29/street-vendors-garbage-pickers-need-protection-climate-extremes/ [Accessed: January 01, 2024]

[22] Salguero-Caparrós MC, Pardo-Ferreira M, Martínez-Rojas JC, Rubio-Romero. Management of legal compliance in occupational health and safety. A literature review. Safety Science. 2020;**121**:111-118

[23] Salamandane A, Malfeito-Ferreira M, Brito L. The socioeconomic factors of street food vending in developing countries and its implications for public health: A systematic review. Food. 2023;**12**(20):3774. DOI: 10.3390/foods12203774

[24] FAO and WHO. Food safety is everyone's business in street food vending. 2022. Available from: https://iris.who.int/bitstream/handle/10665/354468/WHO-HEP-NFS-AFS-2022.4-eng.pdf [Accessed: January 01, 2024]

[25] Khuluse DS, Deen A. Hygiene and safety practices of food vendors. African Journal of Hospitality, Tourism and Leisure. 2020;**9**(4):597-611. DOI: 10.46222/ajhtl.19770720-39 [Accessed: January 01, 2024]

[26] Hueso A. A tale of clean cities: How to solve the urban sanitation challenge.

2016. Available from: https://www.devex.com/news/a-tale-of-clean-cities-how-to-solve-the-urban-sanitation-challenge-88714 [Accessed: January 01, 2024]

[27] Okurut K, Kulabako RN, Abbott P, Adogo JMJ, Pedley S, Tsinda A, et al. Access to improved sanitation facilities in low-income informal settlements of east African cities. Journal of Water, Sanitation and Hygiene for Development. 2015;**5**(1):89-99. DOI: 10.2166/washdev.2014.029

[28] Doda I. Improving the Street Vendors Act: Limitations and Gaps and Way Forward. Durban, Sudáfrica: Streetnet International; 2022. Available from: https://streetnet.org.za/2022/09/09/improving-the-street-vendors-act-limitations-and-gaps-and-way-forward/ [Accessed: January 01, 2024]

[29] Eromosele E. Empowering Local Governments to Improve Access to Clean Water, Sanitation, Hygiene. Lagos, Nigerian: BUSINESSDAY; 2023. Available from: https://businessday.ng/opinion/article/empowering-local-governments-to-improve-access-to-clean-water-sanitation-hygiene/ [Accessed: January 01, 2024]

[30] Rosa-Aquino P. 5 Ways we Need to Adapt to Climate Change — Or Pay the Price. Seatle, WA: Grist; 2019. Available from: https://grist.org/article/5-ways-we-need-to-adapt-to-climate-change-or-pay-the-price/ [Accessed: January 01, 2024]

[31] Matt C. Available at: Disaster preparedness: Build relationships with vendors. 2008. Available from: https://www.facilitiesnet.com/emergencypreparedness/article/Disaster-Preparedness-Build-Relationships-with-Vendors--10008 [Accessed: January 01, 2024]

[32] Recio RB. Street Vendors' Self-Help Strategies Highlight Cities' Neglect of How the Other Half Survive. Braamfontein, South Africa: The Conversation Africa, Inc.; 2019. Available from: https://theconversation.com/street-vendors-self-help-strategies-highlight-cities-neglect-of-how-the-other-half-survive-110283 [Accessed: January 01, 2024]

[33] Women in Informal Employment: Globalizing and Organizing (WIEGO). Street vendors' laws and legal issues in South Africa. 2014. Available from: https://www.wiego.org/resources/street-vendors-laws-and-legal-issues-south-africa [Accessed: January 01, 2024]

[34] Women in Informal Employment: Globalizing and Organizing (WIEGO). Street vendor licensing and permits – reflections from StreetNet International. 2018. Available from: https://www.wiego.org/sites/default/files/publications/files/Horn-WIEGO-Resource-Document-10.pdf [Accessed: January 01, 2024]

[35] Ngcobo N, Plessis D, Fou O. Street traders in South Africa play a vital role: how their rights can be protected. 2022. Available from: https://theconversation.com/street-traders-in-south-africa-play-a-vital-role-how-their-rights-can-be-protected-180886 [Accessed: January 01, 2024]

Chapter 11

A Novel Approach to Ergonomic Risk Analyses

Emin Tarakçi and Emine Can

Abstract

The ergonomics and comfort of employee's health and working conditions are reflected in the efficiency of the work. For this reason, the analysis and evaluation of ergonomic risks in the working environment is of great importance. A novel REBA-FMEA models-based Pythagorean Fuzzy-VIKOR integrated model approach is introduced to assess ergonomic risks. The proposed methodology incorporates PF-VIKOR methodologies based on integrated REBA-FMEA. The following 10 phases comprise the suggested method. A production line case study was conducted. An assessment is conducted on six distinct hazardous occupational positions. The REBA method is used to compute the risk ratings associated with these hazards. The most optimal outcome in the assessment of multi decision-makers with uncertainty in the risk analysis of ergonomic working positions with the novel technique was obtained by computing Pythagorean fuzzy. The novel model overcomes the limitations of traditional methods with the integration of reliability engineering approaches and Pythagorean fuzzy logic. Assessments of ergonomic risks often involve subjective judgments, especially when considering human factors. Different individuals may perceive risks differently, and this subjectivity can introduce variability into the assessment process. The novel method proposed in this study fills the gap in the literature on the subjectivity of decision makers in evaluations.

Keywords: ergonomics, risk analysis, FMEA, PF-VIKOR, REBA

1. Introduction

The pursuit of occupational health and safety in workplaces has led to a heightened emphasis on ergonomic risk analysis. The growing recognition of the impact of poor ergonomics on employee well-being and productivity has prompted researchers to explore innovative methodologies. Among these, the integration of fuzzy logic into the Rapid Entire Body Assessment (REBA) method has emerged as a promising avenue for a nuanced and comprehensive understanding of ergonomic risks. Fuzzy logic, a mathematical approach that handles uncertainty and imprecision, brings a unique dimension to the conventional REBA method by allowing for more flexible and context-aware risk assessments.

In today's workplace, occupational health and safety are top priorities, with a growing emphasis on reducing ergonomic hazards to improve employee well-being and output. A crucial method for determining and resolving risk factors for musculoskeletal injuries and diseases is ergonomic risk analysis. The Rapid Entire Body Assessment (REBA), one of the well-established methods, has become well-known for being a useful instrument for assessing the ergonomic risks connected to different employment. The REBA method, which was first presented by Hignett and McAtamney in 2000, offers a methodical framework for evaluating postural loading and related hazards in work environments [1].

REBA method categorizes tasks based on predefined postural and force criteria, assigning a risk score that correlates with potential musculoskeletal stress. The versatility of REBA has led to its widespread application across various industries, including manufacturing, healthcare, and office environments, underscoring its adaptability to diverse work contexts [1, 2].

Fuzzy logic systems are a prominent modeling approach because they allow for more effective processing of uncertain and complex ergonomic data [3–5]. Fuzzy logic can be used in modeling and analyzing ergonomic factors that involve uncertainty, helping to achieve results that are closer to real-world conditions.

Ergonomics, as a multidisciplinary science, plays a pivotal role in designing work environments that promote optimal human performance while minimizing the risk of musculoskeletal disorders and injuries. The REBA method, a widely utilized tool for ergonomic risk assessment, traditionally relies on precise categorizations and fixed parameters. However, the incorporation of fuzzy logic introduces a degree of adaptability, enabling the assessment of ergonomic risks in situations where conventional methods may fall short.

In the workplace, addressing ergonomic risks is imperative to enhance employee well-being and productivity. Ergonomics, the science of designing work environments to optimize human performance and well-being, has become increasingly crucial in mitigating the adverse effects of repetitive and strenuous tasks [6].

The motivation of this study to contribute valuable insights that can inform both researchers and practitioners in the field, fostering a safer and healthier work environment for diverse industries and occupations.

Conventional ergonomic risk assessment methods frequently concentrate on certain criteria or aspects. By addressing the shortcoming of considering several criteria at once, an integrated model such as Pythagorean Fuzzy-VIKOR may seek to provide a more comprehensive assessment of ergonomic risks. In most ergonomic assessments, dealing with ambiguous and inaccurate information is a requirement. The Pythagorean Fuzzy-VIKOR model may close a gap in the handling of ambiguous data by assisting in the management of uncertainty and vagueness in ergonomic risk assessments. The decision support gap for ergonomic interventions may be filled by the integrated model. REBA-FMEA based integration of PF-VIKOR method has the potential to offer a methodical approach to ergonomic risk management decision-making, facilitating the identification and implementation of interventions.

This study aims to provide a comprehensive overview of the strengths, limitations, and future directions in utilizing REBA for ergonomic risk assessment. The synthesis of existing knowledge will contribute to the ongoing efforts to create safer and healthier work environments across various industries.

2. Methodology

2.1 Rapid entire body assessment (REBA)

The body is divided into two main components using the REBA method, which was first presented by Hignett and McAtamney [1]. The first part is made up of the neck, trunk, and legs. Table A in the REBA worksheet is used to aggregate their scores into a single value. The scores from the upper arm, lower arm, and wrist in the second section are added together using Table B from the REBA worksheet. Table C is used to combine the scores from these tables once the coupling and force scores have been added. The score related to the kind of action is added last.

The final REBA score has a range from one to greater than eleven; the higher the final score is, the greater the hazards of Work-related Musculoskeletal Disorders will be.

Table 1 displays these scores together with the corresponding action levels.

2.2 Failure mode and effects analysis (FMEA)

FMEA is an effective problem-prevention method that complements a variety of engineering and reliability approaches. FMEA has a broad impact on identifying possible product/process failures and planned actions to those failures/hazards, which enhances effective risk management [7].

Failure mode is defined in FMEA as the way in which a product or process can fail. Design faults, mistakes made by people, unpredictable processes, and other unforeseen reasons could be to blame for this. The effects of these failures are related to the possible repercussions of these failures [8].

Risk priority number (RPN) is calculated by multiplying the three parameters "Occurrence", "Severity" and "Detectability" as follows.

$$RPN = [\text{Occurrence}]^{*}\ [\text{Severity}]^{*}\ [\text{Detectability}] \tag{1}$$

Decision makers/experts rate these three parameters on a scale of 1 to 10 based on assessment standards. As it is a measure of the risk of RPN, it can be used to rank errors/hazards and prioritize actions. Actions are taken by prioritizing the error/hazard with the highest RPN.

REBA score	Risk level	Action level
1	Negligible	Not necessary
2–3	Low	It may be necessary
4–7	Middle	Necessary
8–10	High	Necessary soon
11–15	Very high	Needed immediately

Table 1.
REBA risk and action levels [1].

2.3 Pythagorean fuzzy sets and VIKOR (PF-VIKOR)

Among the various extensions of fuzzy set theory, Pythagorean Fuzzy Sets (PFS) have become a powerful tool because they allow for more flexibility and accuracy in capturing the subjective and imprecise information of decision-makers [9, 10].

PFS, initially put forth by Atanassov and then extended by Yager, provide a more expansive framework for handling ambiguity and vagueness by permitting a greater range of membership and non-membership degrees, so easing the limitations imposed by Intuitionistic Fuzzy Sets (IFS).

Opricovic's VIKOR [11] technique offers a methodical and quantifiable approach to multi-criteria decision making, whereas PFS offers a framework for handling ambiguity. It arranges options in order of compromise.

2.3.1 Pythagorean fuzzy sets

First definition: X is a set inside a discourse universe. The shape of a Pythagorean fuzzy set P is as follows [12]:

$$P = \{ <x, P(\mu_P(x), v_P(x))> | x \in X \} \quad (2)$$

where μP(x): X ›→ The membership level is indicated by [0, 1], and vP(x): X ›→ The degree of nonmembership of element x ∈ X to P is represented by [0, 1], and it takes for each x ∈ X.

$$0 \leq \mu_P(x)^2 + v_P(x)^2 \leq 1 \quad (3)$$

For every PF set P and x ∈ X, πP(x) = = $\sqrt{1 - \mu^2 P(x) - v^2 P(x)}$ is the degree of x's indeterminacy with respect to P.

Second definition: Considering two fuzzy Pythagorean numbers P_1 = P(μ_{p1}, v_{p1}) and P_2 = P(μ_{p2}, v_{p2}), and > 0, The operations listed below are defined [12]:

$$P_1 \oplus P_2 = P\left(\sqrt{\mu_{p1}^2 + \mu_{p2}^2 - \mu_{p1}^2\mu_{p2}^2}, v_{p1}v_{p2}\right) \quad (4)$$

$$P_1 \oplus P_2 = P\left(\mu_{p1}\mu_{p2}, \sqrt{v_{p1}^2 + v_{p2}^2 - v_{p1}^2 v_{p2}^2}\right) \quad (5)$$

$$\lambda P_1 = P\left(\sqrt{1 - \left(1 - \mu_{p1}^2\right)^{\lambda}}, (v_{p1})^{\lambda}\right), \lambda > 0 \quad (6)$$

$$P_1^{\lambda} = P\left(\left(\mu_{p1}\right)^{\lambda}, \sqrt{1 - \left(1 - v_{p1}^2\right)^{\lambda}}\right), \lambda > 0 \quad (7)$$

Third definition: Considering two fuzzy Pythagorean numbers P_1 = P(μ_{p1}, v_{p1}) and P_2 = P(μ_{p2}, v_{p2}) the following is the determination of a natural quasi-ordering on the Pythagorean fuzzy numbers [12]:

$$P_1 \geq P_2 \text{ only if } \mu_{p1} \geq \mu_{p2} \text{ and } v_{p1} \leq v_{p2} \quad (8)$$

Ref. [12] provide a scoring mechanism to compare the following two Pythagorean fuzzy numbers of magnitude:

$$s(P_1) = \left(\mu_{p1}\right)^2 - \left(v_{p1}\right)^2 \tag{9}$$

Fourth definition: Using the recommended scoring functions, the following laws [12] are defined for the Pythagorean fuzzy numbers given above in order to compare two of them:

$$\begin{aligned} &\text{If } \mathrm{s}(P_1) < \mathrm{s}(P_2), \text{then P1<P2} \\ &\text{If } \mathrm{s}(P_1) > \mathrm{s}(P_2), \text{then P1>P2} \\ &\text{If } \mathrm{s}(P_1) = \mathrm{s}(P_2), \text{then P1} \sim \text{P2} \end{aligned} \tag{10}$$

2.4 The suggested integrated REBA-FMEA based PF-VIKOR method

Integrated REBA-FMEA based PF-VIKOR approaches are incorporated into the suggested methodology. The suggested method is broken down into the following ten steps:

Step 1: Define hazardous working positions and ergonomic risks.

Step 2: Calculate the REBA score of these risks.

Step 3: Matching the REBA score with the severity parameter of FMEA according to a seven-point Pythagorean fuzzy linguistic scale.

Step 4: Determine weight of occurrence and detectability parameters.

Step 5: Determine weight of decision makers.

Step 6: Evaluate the occurrence and detectability parameters by decision makers with 7 PF linguistic terms.

Step 7: The VIKOR S and R values should be calculated using VIKOR.

Step 8: Compute VIKOR Q values.

Step 9. Using the S, R, and Q values as a guide, order the risk priority numbers.

Step 10: Based on the risk priority number, decide on control and preventative measures.

3. Case study

A case study was carried out on the production line considered in the study of Tarakçı [6]. The study population, hazardous working positions and REBA scores were expanded with reference to this study [6].

For step 1 and step 2, the data in Tarakçı's study [6] are taken as reference. Six different hazardous working positions are assessed. Risk scores of these hazards are calculated with the REBA method.

For step 3, the matching of the REBA score to the severity parameter of the FMEA according to the seven-point Pythagorean fuzzy linguistic scale is as shown in **Table 2** below.

For Step 4, the weights of the occurrence and detectability parameters are Wo = 0.3641, Ws = 0,3369, and Wd = 0.2989 obtained in Tarakçı's PhD thesis [13].

For step 5, teams representing occupational health and safety (DM1), production line workers (DM2), and managers (DM3) made composed the decision-making committee.

For every decision maker, there were several weights assigned. Assigning decision makers is done by the computational method outlined in [14]. A weight (wdm > 0 and $\sum$wdm = 1) is assigned to each decision maker. The allocated weights are, in order, wDM1 = 0.5, wDM2 = 0.3, and wDM3 = 0.2.

REBA score	Meaning at PF	Fuzzy numbers (u,v)
1	Very Low (VL)	(0.15,0.85)
2–3	Low (L)	(0.25,0.75)
4–5	Moderately Low (ML)	(0.35,0.65)
6–7	Medium (M)	(0.50,0.45)
8–9	Moderately High (MH)	(0.65,0.35)
10–12	High (H)	(0.75,0.25)
13–15	Very High (VH)	(0.85,0.15)

Table 2.
Matching table of REBA score and severity parameter according to PF linguistic scale.

Meaning	Corresponding Pythagorean fuzzy number (u, v)
Very Low (VL)	(0.15,0.85)
Low (L)	(0.25,0.75)
Moderately Low (ML)	(0.35,0.65)
Medium (M)	(0.50,0.45)
Moderately High (MH)	(0.65,0.35)
High (H)	(0.75,0.25)
Very High (VH)	(0.85,0.15)

Table 3.
The seven-point Pythagorean fuzzy linguistic scale [15].

For Step 6, decision makers rated the occurrence and detectability parameters with seven-point Pythagorean fuzzy logic linguistic terms as shown below (**Table 3**).

3.1 Case result

S, R, and Q values are found for Steps 7, 8, and 9. The formulation sets were computed using PyCharm Community program [16].

Table 4 shows the Q values and risk priority numbers for 6 hazardous working positions. Risk priority numbers are ranked according to Q values. The lowest Q value

Hazard working position	Rank	Q value
HWP1	6	0.986
HWP2	1	0.0
HWP3	3	0.804
HWP4	2	0.382
HWP5	4	0.845
HWP6	5	0.959

Table 4.
Q values and ranking orders for hazardous working positions.

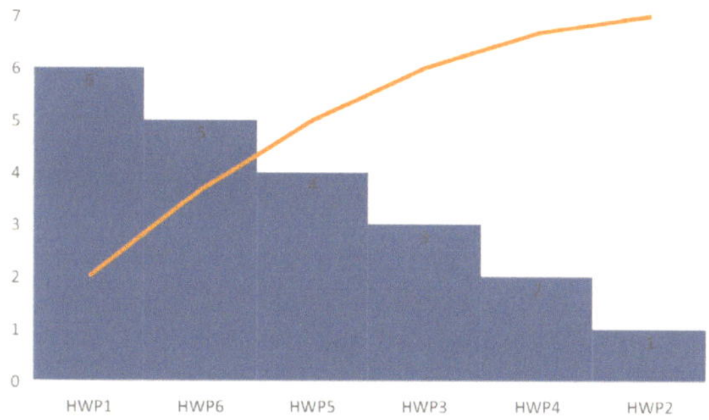

Figure 1.
Sequencing on the Q value curve.

indicates the most important risk. The highest Q value is the risk with the lowest priority score.

As can be seen from **Table 4** and **Figure 1**, the three most important and prioritized hazardous working positions are HWP2, HWP4 and HWP3 respectively. In addition, the least hazardous working positions are HWP1, HWP6 and HWP5 respectively.

4. Conclusion

In conclusion, the integration of Pythagorean fuzzy with the REBA method has proven to be a powerful approach in the analysis and evaluation of ergonomic risks within various workplace settings. The amalgamation of these two methodologies addresses the inherent uncertainties and complexities associated with ergonomic data, providing a more nuanced and adaptable framework for risk assessment.

The application of fuzzy logic allows for a flexible representation of imprecise and uncertain ergonomic variables, which are prevalent in real-world occupational environments. By incorporating Pythagorean fuzzy into the REBA method, the analysis gains the capability to handle ambiguous data and account for the vagueness inherent in human postures and movements.

The findings of this study highlight the significance of utilizing a comprehensive approach that combines the precision of the REBA method with the adaptability of Pythagorean fuzzy logic. The integrated model not only facilitates a more accurate assessment of ergonomic risks but also enhances the ability to prioritize and manage these risks effectively.

Moreover, the outcomes of the analysis contribute to the broader field of ergonomics by providing insights into the dynamic nature of occupational risk factors. The adaptability of Pythagorean fuzzy logic-integrated REBA method ensures that it can be applied across diverse industries and work scenarios, making it a valuable tool for researchers, practitioners, and organizations seeking to optimize workplace conditions.

While this study showcases the potential of the integrated approach, it is essential to acknowledge its limitations. Further research could explore refinements in the fuzzy logic model and the validation of the integrated method in various industrial contexts.

This study demonstrates that the Pythagorean fuzzy-VIKOR-based novel scoring system was more sensitive to changes in input variables than the conventional approaches. When assessing MSDs, one might apply this proposed model.

The PF-VIKOR model could help handle ambiguity and vagueness in ergonomic risk assessments, which could fill a gap in the handling of confusing data. The integrated model may close the gap in decision support for ergonomic interventions. REBA-FMEA based integration PF-VIKOR method has the potential to provide a systematic approach to ergonomic risk management decision-making, making the identification and execution of interventions easier.

In conclusion, Pythagorean fuzzy-integrated REBA method offers a promising avenue for advancing ergonomic risk analysis. The synergy between the precision of REBA and the flexibility of fuzzy logic provides a robust foundation for future developments in optimizing workplace environments, promoting employee well-being, and fostering sustained productivity.

Conflict of interest

No conflict of interest was declared by the authors.

Research limitations

The research is limited to the opinions and assessments of experts (decision makers) and the hazardous working positions and space/time dimensions addressed in the study.

Author details

Emin Tarakçi[1*] and Emine Can[2]

1 ULAK Haberlesme Inc., Istanbul, Turkey

2 Istanbul Medeniyet University, Istanbul, Turkey

*Address all correspondence to: tarmuhendislik@gmail.com

References

[1] Hignett S, McAtamney L. Rapid entire body assessment (REBA). Applied Ergonomics. 2000;**31**(2):201-205. DOI: 10.1016/S0003-6870(99)00039-3

[2] Harrison AJ, Kimani PK. The application of the rapid entire body assessment (REBA) and the strain index (SI) in the assessment of risk factors for work-related musculoskeletal disorders in a furniture manufacturing company. Ergonomics. 2006;**49**(6):544-556. DOI: 10.1080/00140130500520485

[3] Zhang X, Chaffin DB. Digital Human Modeling for Computer-Aided Ergonomics. USA: Interventions, Controls, and Applications in Occupational Ergonomics; 2005

[4] Hatiboglu M, Dayioglu H, İssever H, Ayvaz B. Selection of ergonomic risk assessment method with pythagorean fuzzy sets: Practice in emergency medical services. Journal of Intelligent & Fuzzy Systems. 2023;**45**(1):391-405

[5] Falahati M, Dehghani F, Malakoutikhah M, Karimi A, Zare A, Yazdani Rad S. Using fuzzy logic approach to predict work-related musculoskeletal disorders among automotive assembly workers. Medical Journal of the Islamic Republic of Iran. 2019;**33**:136. DOI: 10.34171/mjiri.33.136

[6] Tarakçı E, Can E, Sakallı AE, Tak G. The ergonomic risk analysis with REBA method in production line. Ergonomi. 2020;**3**(2):96-107. DOI: 10.33439/ergonomi.743276

[7] Ireson G, Coombs W, Clyde F, Richard YM. Handbook of Reliability Engineering and Management. 2nd ed. New York, NY: McGraw-Hill Professional; 1995

[8] Stamatis DH. Failure Mode and Effect Analysis: FMEA from Theory to Execution. Milwaukee, USA: ASQ Quality Press; 2003

[9] Atanassov KT. New operations defined over the intuitionistic fuzzy sets. Fuzzy Sets and Systems. 1994;**61**(2):137-142

[10] Yager RR. Pythagorean membership grades in multicriteria decision making. IEEE Transactions on Fuzzy Systems. 2013;**22**(4):958-965

[11] Opricovic S, Tzeng G-H. Compromise solution by MCDM methods: A comparative analysis of VIKOR and TOPSIS. European Journal of Operational Research. 2004;**156**(2):445-455

[12] Zhang X, Xu Z. Extension of TOPSIS to multiple criteria decision making with Pythagorean fuzzy sets. International Journal of Intelligent Systems. 2014;**29**(12):1061-1078

[13] Tarakçı E. Güvenlik mühendisliği yaklaşımı ile pandemi döneminde risk analizi ve yönetimi: Sağlık kuruluşlarında Covid-19 virüsünün bulaş ve yayılma süreçlerinin incelenmesi [Doctoral dissertation]. University Library. Istanbul, Turkey: Istanbul Medeniyet University; 2023

[14] Kabir S, Yazdi M, Aizpurua JI, Papadopoulos Y. Uncertainty-aware dynamic reliability analysis framework for complex systems. IEEE Access. 2018:29499-29515. DOI: 10.1109/ACCESS.2018.2843166

[15] Cui FB, You XY, Shi H, Liu HC. Optimal siting of electric vehicle charging stations using Pythagorean fuzzy VIKOR approach. Mathematical Problems in Engineering. 2018:32-44

[16] PyCharm Community Edition. (Version 2023.1.1) [Computer software]. 2023